Acidity Management in Musts and Wines

Second Edition

Acidification, deacidification, crystal stabilization, and sensory consequences

Volker Schneider

Sarah Troxell

WINE APPRECIATION GUILD PRESS

SAN FRANCISCO

Acidity Management in Musts and Wines
Second Edition

Text copyright © 2018 , 2021 Volker Schneider, Sarah Troxell

Wine Appreciation Guild Press
an imprint of
Board and Bench Publishing
www.boardandbench.com

Editorial direction by Annika LD Imelli
Design and composition by PerfecType, Nashville, TN

ISBN
978-1-935879-25-1

Library of Congress Cataloging-in-publication is on file with the Library of Congress

Preface

This is a kind of book no enologist is keen to write. It is about corrections in winemaking, more precisely about acidity adjustment. As a growing number of today's winemakers are embracing the expression of terroir, non-interventionist winemaking, and the concept that serious wine is made exclusively in the vineyard, any kind of correction is being brought into increasing disgrace. However, acidity adjustments are one of the most common interventions in winemaking, disclosing the shortcomings of viticultural measures, guaranteeing harmoniously balanced acidity in the finished wine. The typical tartaric acid additions to musts resulting from hot-climate growing conditions are a prime example of that, while winemakers in cool-climate conditions more often feel the need for lowering acidity. Thus, acidity management and crystal stabilization associated therewith are key issues in global winemaking.

In this context, the pH value is an important and commonly discussed analytical parameter, but an ideal pH alone does not guarantee sensory quality and wine shelf life. The sour taste is one of the most important elements in the sensory perception of wine. Fractions of 1 g/L of titratable acidity determine whether a wine is perceived as harmonious, too sour, or too flat. However, acidity corrections comprise much more than a mere numerical shift of titratable acidity and pH. Potassium also plays an important role. Sensory perception of sourness is not that strongly correlated with these metrics as one might expect, just as all in-mouth perceptions result from the interaction of a broad number of variables.

As a result, many wines are deemed either too harsh and sharp or too flabby and soapy. Questions arise and continue to be a topic of technical discussions - how much of which kind of acid to add for acidification, which kind of carbonate to use for deacidification, and how to achieve crystal stability. Hence, practitioners and trainees feel the understandable need of an in-depth treatise on these issues.

This booklet aims at filling the gap. It does not claim to be an academic one dealing with hard core science, which can be consulted in the literature cited. It is not about innovative enological concepts, not about ground-breaking revelations on how to make the best wine, but just about sober facts governing the sensory outcome of one of the most important tools in winemaking. It has a strong sensory focus and provides numerous hands-on instructions on how to conduct sensory trials. These trials can be used by students for guidance in workshops and by winemakers as examples of bench trials for sensory optimization by acidity adjustment.

The authors accept liability for any errors contained in this book. Their disclosure will be taken into account in future editions.

Table of contents

1.

Chemical basics and definitions

This first chapter provides a brief introduction to the chemical basis of acidity corrections and defines the technical terms used in this context. It deals with concepts such as titratable acidity, total acidity, pH, buffer capacity, ash alkalinity and their measurement as well as with the strength of individual acids and the role of mineral cations contained in wine. This is not complicated chemistry, but concepts that many have learned in school and college and have since forgotten. Without any doubt the most theoretical part of this book, this chapter can be skipped by readers who have a strong grasp of chemistry and might start with chapter 2.

1.1. Total acidity and titratable acidity

Seen from the angle of acidity management and crystal stabilization, musts and wines can be considered as an aqueous or hydroalcoholic solution of acids and bases. By its nature, this solution is contaminated by other molecules, especially those of a colloidal nature, which makes the expected precipitation of poorly soluble salts of acids and mineral cations difficult or impossible.

Acids release hydrogen (H^+) ions, also called protons, into the solution. The H^+ ions turn the solution acidic. The remaining acid residue is an anion bearing a negative charge. Bases release hydroxide (OH^-) ions into the solution. The remaining base residue is a cation bearing a positive charge. It neutralizes acid residues converting them into salts.

The acids can be present in a free, a partially neutralized, or a totally neutralized form. In musts and wines, their prevailing part consists of organic acids and a minor fraction of inorganic acids. The most important organic acids comprise tartaric, malic, acetic, and lactic acid. The inorganic acids encompass essentially minor amounts of phosphoric, sulfuric, sulfurous, and sometimes ascorbic acid.

The cations that are responsible for the partial neutralization of acids are predominantly of inorganic, mineral nature. The most important one of them is potassium, accompanied by calcium, magnesium, and sodium at distinctly lower concentrations. These mineral cations neutralize preferentially the strongest ac-ids such that sulfuric acid is completely and phosphoric acid partially neutral-ized. The less strong organic acids are, as a matter of principle, only partially neutralized. It should be recalled in this context that strong acids are those which dissociate easily or even completely into their ions in water, whilst weak acids do so to a lesser extent.

Neutralization converts the acids into their corresponding salts. Therefore, in the enological context, sulfuric acid is also referred to as sulfate, phosphoric acid as phosphate, tartaric acid as tartrate, etc.

In wine, H^+ ions are present in excess since wine contains more acids than its mineral cations are able to neutralize. Therefore, wine is an acidic solution. The sum of all free, non-neutralized acids is measured by titration using standard alkali solutions of sodium hydroxide (NaOH) or potassium hydroxide (KOH) and designated as titratable acidity (TA). Or more precisely in chemical terms: Titratable acidity is measured as the sum of H^+ ions the undissociated organic acids are able to dissociate plus free H^+ ions already dissociated.

Titratable acidity is usually expressed as tartaric acid equivalents. Its measure-ment comprises only the non-neutralized portions of tartaric acid, malic acid, lactic acid, etc. Figure 1 illustrates this relationship.

Figure 1: The quantitative relationship between free acids, salts, and mineral cations.

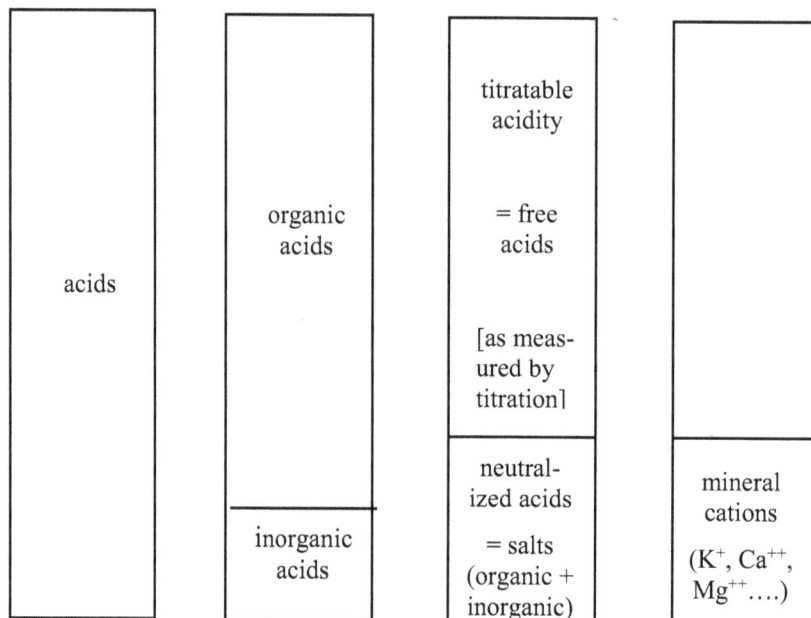

There are four principal metrics associated with acidity that are used to describe the acidic features of wine. They comprise:

– total acidity

– titratable acidity

– pH

– buffer capacity

Titratable acidity and pH are the most important of these in wine quality control.

Titratable acidity as a part of total acidity

Total and titratable acidity are terms that are frequently misused and exchanged by wine practitioners, although they have a quite different meaning.

If all individual acids in a wine are determined by specific chromatographic, enzymatic, or spectrophotometric analysis, expressed as tartaric acid or other equivalents and summed up, the total will be greater than the titratable acidity concentration. This is because the measurement of individual acids records the concentration of their respective anions irrespective of whether they are totally neutralized, partially neutralized, or actually occurring as free acids. In contrast, when acidity is measured by direct titration of the wine with a base, the result represents only the sum of the free portions of all acids. Therefore, titratable acidity is always lower than the sum of the individual acids constituting it.

As an example, let's consider a solution of potassium bitartrate. It is a salt of tartaric acid neutralized to 50 % by potassium. Let's further consider that the initial concentration of tartaric acid in that solution was 1.0 g/L. If the solution is analyzed for titratable acidity by titration, the result will be only 0.5 g/L expressed as tartaric acid. However, when the same solution is analyzed for total acidity, using a spectrophotometric method for example, the result will be 1.0 g/L.

Instead of quantifying and summing individual acids, total acidity can more easily be measured by treating the wine with a strong cation exchange resin in the H^+ form prior to titration. The cation exchanger removes mineral cations responsible for the partial neutralization of acids by exchanging them for H^+ ions. The concentration difference obtained by titration before and after this treatment is equal to the acidity fraction that has been neutralized, or equivalent to the mineral cations concentration.

The acidity as measured by plain titration on untreated wines corresponds to only some 75 % of the total acidity obtained after cation exchange treatment or by adding up the individual acids measured specifically (Boulton 1980 a). In other words, 25 % of the acids are neutralized on average. The exact extent depends on the concentration of mineral cations available to neutralize them. Owing to this difference, the "total acidity" measured as a routine analysis in the wine industry is more and more replaced by the more precise terms of "titratable acidity" or "total titratable acidity". Hereinafter, the term "titratable acidity" is used and abbreviated as "TA".

The following definitions are essential:

– Titratable acidity is measured as the total H^+ ion concentration, dissociated and not dissociated, from all acids present. It is essential for taste, balance, and other in-mouth sensations.

– pH only measures the H^+ ions that are already dissociated (Chapter 1.2).

– Total acidity measures the concentration of all acids plus their salts.

Different ways of expressing titratable acidity

After establishing the definition of titratable acidity (TA), there is still need of further discussion. This is because TA as measured routinely in any winery lacks a uniform way of expressing its concentration.

Theoretically, TA can be expressed as a level of acidity in terms of any acid. Since tartaric acid is the primary acid in grapes and winemaking, it has become the reporting acid for the measurement in almost all wine growing countries. However, reporting titratable acidity as tartaric acid and writing "the wine contains 6.5 g/L as tartaric acid" does not mean the wine actually contains 6.5 g/L of tartaric acid. Rather, it means the wine has the same titratable acidity concentration as a 6.5 g/L tartaric acid solution.

France is an exception to this rule because it reports TA as sulfuric acid equivalents. The conversion factor between both units is 1.53, i.e., 1.0 g/L sulfuric acid equals 1.53 g/L tartaric acid. This means that when a wine has been reported by a French lab as 4.5 g/L TA, it would be analyzed as 6.9 g/L TA in any other country. Thus, reporting TA without specifying the reference acid is not helpful.

Expressing TA in a more correct way as mEq/L (milliequivalents per liter) has not been embraced by the wine industry.

A further problem in comparing TA data stems from different titration end points used in this simple acid-base titration. European countries titrate grape wine TA to pH 7.0 endpoint, according to OIV recommendations. pH 7.0 means that the solution is neutral, and that H^+ and OH^- ions are present at an equal ratio. However, pH 7.0 as the endpoint of TA titration in wine has been predefined arbitrarily. It does not allow for assessing the entirety of free acids present in wine since it does not take into account the true pH at which they are completely neutralized. Therefore, most other countries including North America use pH 8.2 as titration endpoint for TA measurements according to AOAC standards. For a deeper understanding of the difference, a short excursion into the theoretical basics of acid-base titrations is useful.

Neutral point is not equivalence point

When an acid mix in a solution such as TA of wine is measured by titration, a standard solution of a strong base is used as titrant and added stepwise to an accurately measured volume of sample until the amount of base equals the amount of the acid to be quantified. At this moment during titration, a steep increase of pH can be observed. This pH leap corresponds to the equivalence point. The equivalence point is the pH where the acid and the base used for its titration are present in equivalent amounts.

Plotting pH against the volume of added base yields the titration curve. A typical titration curve obtained on grape wine is depicted in figure 2.

Figure 2: Titration curve of a white wine.
* = equivalence point

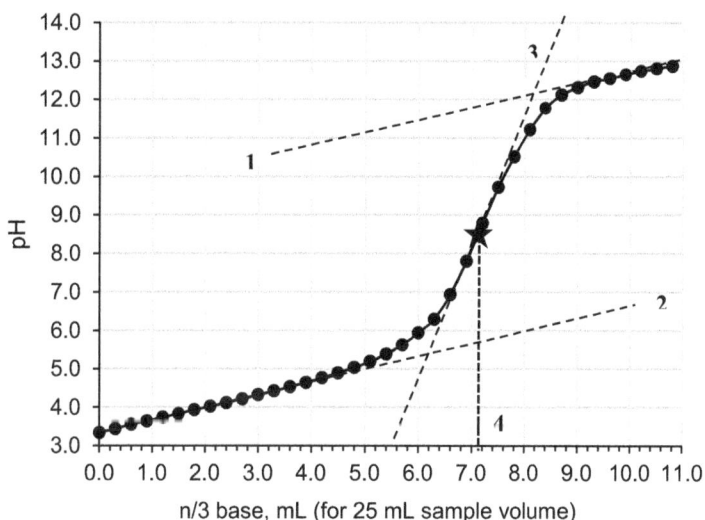

n/3 base, mL (for 25 mL sample volume)

The slope of a titration curve depends on whether one measures the concentration of a strong acid or a weak acid. Titration of a strong acid yields an equivalence point, which corresponds to the neutral point at pH 7.0. However, when weak acids or the acid mixture of wine are titrated with a strong base (NaOH or KOH) until complete neutralization, the equivalence point does not coincide with the neutral point, but shifts to the alkaline area of pH > 7.0.

It is worth noting that the weaker the acids, the more the equivalence point shifts towards higher pH and the less the pH jumps around the equivalence point. The reason is that the anions of weak acids act themselves as bases of the corresponding acids, reacting with water and releasing OH⁻ ions. Therefore, OH⁻ ions prevail in solution at the equivalence point and turn it alkaline, i.e. pH is > 7.0. In other terms, the salts of weak acids display a slightly alkaline reaction. Consequently, the equivalence point does not coincide with the neutral point.

The titration curves of wines correspond basically to those of weak acids, but they vary slightly between wines according to their titratable acidity, the concentrations and strength of their individual acids, and their mineral cation content. They display almost linear behavior up to pH 5 before they start to surge upwards very steeply around a pH corresponding to the equivalence point. At this pH, neutralization of organic acids is completed; the curve slope is at its maximum and shows an inflection point. The inflection or equivalence point is calculated by the 1st derivative method or determined by the geometric method. Modern titration analyzers calculate it automatically.

The geometric method draws two lines (lines 1 and 2 in figure 2) that follow the flat, more horizontal parts of the curve before and beyond the inflection point each. A third line (line 3) follows the steep, more vertical part of the curve. The half-way distance between the top and the bottom intersection of the third line is the equivalence point. A fourth line (line 4) drawn vertically from the equivalence point to the x-axis gives the titrant volume at the equivalence point.

The titration curve in figure 2 shows an equivalence point at pH 8.5 using the geometric method. This is the pH where all acids contained in the wine sample have been completely neutralized by titration. When the titration is carried out to this pH, titratable acidity (TA) will be measured as 7.2 g/L and represent its true content.

According to AOAC rules, TA is only titrated to pH 8.2 endpoint, which results in approximately 7.0 g/L TA considered as the accurate value in this example. On the other hand, the OIV method used in Europe with a pH 7.0 endpoint would yield a lower 6.6 g/L TA. Remember, all TA concentrations obtained for this wine are expressed in tartaric acid as the reference.

In a large number of grape must and wine samples, equivalence points have been shown to vary between pH 7.5 and 8.5 with an average of pH 8.2 (Schaller and Paul 1958, 1959, Guymon and Ough 1962). Thus, titrating TA to pH 8.2 endpoint, though arbitrarily decreed for all wines, yields results very close to reality. For standard wines, these results are approximately 0.5 g/L higher than those obtained by the OIV method used in Europe (Wong and Caputi 1966). There is a fairly good correlation between both, which has been described in statistical terms (Darias-Martin et al. 2003) provided that they are all expressed in tartaric acid as a common reference acid.

Typical TA levels are 5 – 8 g/L in white wines and 4 – 6 g/L in red wines (as tartaric acid to pH 8.2).

Corrections for carbon dioxide (CO_2) and sulfur dioxide (SO_2)

CO_2: Despite the differences in expressing wine TA, there is consensus that it should not include the carbon dioxide (CO_2) contained in wines. At wines' low pH, CO_2 exists as a dissolved gas but is converted into carbonate when pH increases during TA titration and therefore measured as an acid. Thus, high CO_2 levels in young wines after fermentation may feign 0.5 to 1.0 g/L more TA than the wine actually contains. The false TA readings caused by CO_2 are even higher in sparkling wines. For that reason, CO_2 has to be carefully removed before proceeding to TA titration. Degassing the sample can be achieved by holding it at approximately 70° C in a water bath during several hours, by using a vacuum pump, or more easily by vigorous shaking at ambient temperature.

The primitive looking but effective shaking method is helpful in many winery labs, when a small volume of wine is repeatedly shaken in a thumb-sealed bottle. When pressure release is no longer felt when the thumb is raised, the wine is essentially free of CO_2. In this case, the pH remains stable at the titration end point. However, if the sample still contains too much CO_2, the pH will gradually drop again after reaching the titration end point. This phenomenon is due to the formation of carbonates from the CO_2. If titration continues in such a case and base is added again, too high values are obtained for the titratable acidity because CO_2 is also included.

SO_2: Sulfur dioxide (SO_2) is also not considered part of TA legally. 1.0 g/L of total SO_2 increases TA readings by 2.34 g/L (as tartaric acid). However, its low concentrations in today's' wines minimizes the analytical error in TA titrations. Therefore, SO_2 impact on TA readings is usually neglected. On the other hand, the inaccuracy becomes inadmissibly high when juices are stored with high amounts of SO_2 exceeding several hundred mg/L to prevent fermentation. Removal of SO_2 can then be achieved by boiling the sample in an open vessel for several minutes and subsequent volume adjustment.

Fixed and volatile acidity

TA includes both volatile and non-volatile acidity. Volatile acidity (VA), mainly composed of acetic acid, is considered to be a part of TA since it contributes to sourness. In quality control, it is often determined separately and expressed as acetic acid. After conversion of acetic acid into tartaric acid units, subtracting VA from TA yields the fixed acidity.

Alkalinity and ash alkalinity

Likewise the sum of the non-neutralized fractions of the organic acids is measured as TA, the individual mineral cations can be expressed as a sum parameter in milliequivalents per liter (mEq/L). This value is called alkalinity. More frequent is the assessment of ash alkalinity. To understand this term, one must first have an understanding of what ash is.

Ash equals the sum of all inorganic substances of wine, which remain after incineration of its evaporation residue. It includes all mineral cations (potassium, calcium, magnesium, sodium....) and inorganic anions (phosphate, sulfate...), which are absorbed during grape ripening from the soil. Its content in wine varies between 1.5 and 3.5 g/L. It can be determined gravimetrically after incineration of wine extracts and complete combustion of carbon at 500-550° C, or calculated based on the individual measurements of cations and anions.

The ash alkalinity corresponds to the sum of the cations, other than the minor amount of ammonium cation, which neutralize organic acids. Its measurement involves titration of the ash, which is accomplished by adding acid to the ash until the solution is neutralized. Results are expressed in milliequivalents per liter.

In current practice of wine quality control, ash and ash alkalinity do not play a major role. Their measurement is usually replaced by the determination of individual mineral cations such as potassium or calcium, which can provide important information in commercial winery settings, depending on the wine and the question to be answered.

1.2. pH - the hydrogen ion concentration

In most bottled wines, pH is in the range between 3.0 and 4.0, with red wines tending to display values in the upper part of this range. However, when pH exceeds 3.8, their aging potential and shelf life tend to be compromised for chemical reasons regardless of microbial stability achieved by sterile bottling (Schneider and Tracey 2021). pH values lower than 3.0 are indicative of an excessive tartaric acid addition. The reason for this behavior is explained in chapters 2.3 and 3.4.

It is commonly known that the pH of wine or must plays a critical role in the wine industry. It influences the ability of most bacteria to grow, the solubility of tartrate salts, the effectiveness of sulfur dioxide, enzyme, and bentonite additions, the polymerization of red wine tannin and color pigments, and oxidative and browning reactions. In contrast, TA has no direct effect on chemical or enzymatic reactions or microbial activity.

Relatively small numerical differences in pH can become of great importance. It is precisely for this reason that the issue of pH and its interpretation is, despite its triviality, addressed in some detail.

In an aqueous solution, acids release hydrogen (H^+) ions (protons) to the solvent and make it acid (Chapter 1.1). This behavior is called dissociation. In the case of most organic acids, it is the hydrogen of their carboxyl groups (-COOH) that dissociates according to the general formula

$$R-COOH \rightarrow R-COO^- + H^+$$

The readiness of acids to dissociate is variable (Chapter 1.4). The prevailing acids in must and wine, tartaric and malic acid, contain two carboxyl groups, i.e., they are divalent. Therefore, they can at best release two H^+ ions (protons), which is the reason why they are also designated as diprotic. However, as they are relatively weak acids, they only release a small portion of them in the typical pH-range of wine.

In contrast, bases release hydroxyl (OH^-) ions to the solvent or bind H^+ ions themselves. In any solvent, H^+ ions and OH^- ions are present simultaneously, albeit at highly different concentrations in most cases. In an acid solution such as wine, H^+ ions are dominant. The degree of acidity of such a solution stems from its actual concentration of H^+ ions, expressed in mol/L. This leads to confusingly high decimal numbers.

The pH scale

In order to make working with these numbers easier, the molar concentration of H^+ ions is given as pH values. They express the negative decimal logarithm of the molar H^+ concentration:

$$pH = -\log_{10} [H^+]$$

Thus, a solution containing 0.001 mol/L H^+ ions displays pH 3.0, another solution containing 0.0001 mol/L H^+ ions displays pH 4.0, etc.

Table 1 gives the well-known pH scale ranging from 0 to 14. The lower the pH on this scale, the higher the concentration of H^+ ions and the stronger the acidic reaction of the solution. At pH 7.0, H^+ ions and OH^- ions are present at equal concentrations; the solution is neutral. At pH lower than 7.0, the solution is acidic because there are more H^+ ions than OH^- ions.

Table 1: The pH scale

pH	Concentration of H^+-ions (mol/L)		Concentration of OH^--ions (mol/L)	
0	1^0	1,0	0,00000000000001	1^{-14}
1	1^1	0,1	0,0000000000001	1^{-13}
2	1^{-2}	0,01	0,000000000001	1^{-12}
3	1^{-3}	0,001	0,00000000001	1^{-11}
4	1^{-4}	0,0001	0,0000000001	1^{-10}
5	1^{-5}	0,00001	0,000000001	1^{-9}
6	1^{-6}	0,000001	0,00000001	1^{-8}
7	1^{-7}	0,0000001	0,0000001	1^{-7}
8	1^{-8}	0,00000001	0,000001	1^{-6}
9	1^{-9}	0,000000001	0,00001	1^{-5}
10	1^{-10}	0,0000000001	0,0001	1^{-4}
11	1^{-11}	0,00000000001	0,001	1^{-3}
12	1^{-12}	0,000000000001	0,01	1^{-2}
13	1^{-13}	0,0000000000001	0,1	1^{-1}
14	1^{-14}	0,00000000000001	1,0	1^0

Since pH represents a negative logarithmical figure, the pH scale is counterintuitive. Thus, a wine with pH 3.0 contains ten times more H^+ ions than a wine with pH 4.0. Therefore, apparently small differences in pH, for example between pH 3.1 and pH 3.4, become important since there are twice as much H^+ ions at pH 3.1 than at pH 3.4.

In conclusion, pH measurements indicate the actual concentration of free H^+ ions that are actually released by the dissociation of acids under the specific conditions of a given wine. In contrast, TA measurements yield the actually existing, free H^+ ions plus the H^+ ions the acids are potentially able to release until they are neutralized by titration.

In the course of titration, one slowly adds a base whose OH^- ions bind with the H^+ ions released by the acids. The same reaction occurs on the palate when acids are neutralized by mineral cations of the saliva. This means, the sensory cells of the oral cavity not only respond to the actual concentration of H^+ ions, but also to the pH changes elicited by the neutralization the saliva. This explains why titratable acidity correlates relatively well with the sour taste.

The importance of pH in wine

Whenever TA is measured by potentiometric titration, pH readings are obtained before the beginning of the operation. Therefore, it is useful and reasonable to report and interpret them one in context with the other.

Depending on the winemaker's educational background, pH can be either neglected or over-intellectualized. In search of microbial security, some winemakers build their winemaking strategy on pH regardless of the sensory outcome, as though numbers were more important than taste. However, pH has not only stability but also sensory implications.

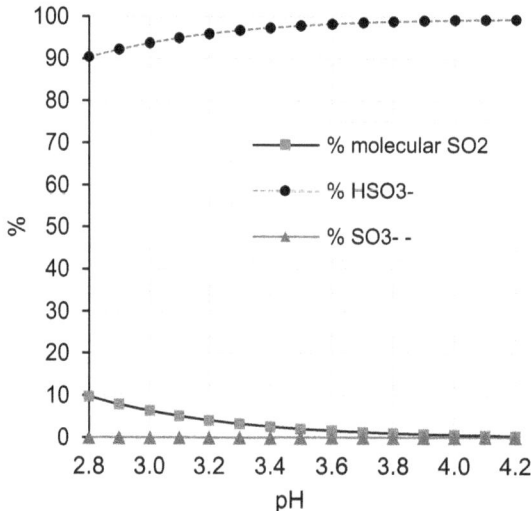

Figure 3: Effect of wine pH on the relative proportions of forms of free sulfur dioxide.

One of the reasons pH is important for wine preservation is the way it determines the microbially active percentage of free SO_2. This fraction is the part of free SO_2 that is present in its undissociated, molecular form as a dissolved gas. The higher the pH, the less free SO_2 will be in the useful molecular form. Figure 3 depicts this relationship.

In order to effectively suppress microbial activity, a minimum of 0.6 to 0.8 mg/L molecular SO_2 is considered necessary. Based on free SO_2 and pH, it can be calculated according to the formula

$$SO_2 \text{ molecular (mg/L)} = \text{free } SO_2 : [1 + 10^{(pH - 1.81)}]$$

The subtrahend 1.81 contained in the power corresponds to the acid constant (pK_A value) of sulfurous acid (Table 2).

Sensory implications

Since low pH is of interest for microbial stability of juice and wine, some wine-makers use pH as a criterion for determining the ideal moment to harvest. When grapes ripen, pH increases because acids are lowered and mineral cations taken up. On the other hand, premature harvest with the objective of achieving a low pH or maintaining a satisfactory level of TA poses a serious risk of unripe aromatics or tannins in the future wine. While it is quite impossible to transform green-vegetal aromatics into ripe ones by enological means, it is easy to correct pH and TA by means of acidity management.

Besides its importance on microbial stability, pH also has a significant sensory impact. It influences the color hue of red wines. In combination with TA, it impacts the sensory perception of sourness. Since it determines the molecular, gaseous portion of free SO_2 that can be perceived by the olfactory system, it also influences the perception of a given level of free SO_2 by smell. This is the reason why a current level of 35 mg/L of free SO_2 in a wine with pH 3.10 can be twice as pungent by smell as in a wine with pH 3.40. Hence, the aroma perception of low pH wines can be seriously biased when free SO_2 adjustments before bottling do not take into account this pH effect. Sensory interferences by SO_2 are much more than just an issue of free SO_2 levels.

The balance of pH and TA

The actual pH of wine stems from an interaction of acids and mineral cations (Chapter 1.5). Chapter 2.3 explains how in connection with TA, pH also allows for a rough estimation of the potassium content, which is of considerable importance for the style and body of wine.

Winemakers are interested in keeping pH low to minimize the risk of microbial spoilage, but also in preventing TA from getting too high to avoid excessive sourness. As will be explained in chapters 3.4 and 3.5, this objective is not always easy to achieve.

TA and pH are inversely correlated, but that correlation is much less perfect than one might expect for two reasons:

– Organic acids vary by their ability to release protons (H^+). This ability is expressed by their pK_A values (Chapter 1.4). Stronger acids result in a greater pH decrease for the same contribution to TA than weak acids.

– At a given TA, higher concentrations of mineral cations, primarily potassium, cause a higher pH. This will be shown in section 2.3 and in figure 8 in particular.

Despite the relationship between titratable acidity and pH, it is difficult to forecast the exact value that pH will reach after a given modification of the acid-cation equilibrium (Figure 1) of a wine. In other words, the value of titratable acidity alone as well as its theoretical variations generated by measures of acidification or deacidification do not allow us to exactly predict the changes of wine pH these variations will determine. Since pH arises out of the interaction between acids and mineral cations, the same pH can be measured in different wines with either low or high TA, and vice versa. Due to the bad correlation between TA and pH, both figures are usually measured and reported simultaneously when titratable acidity is evaluated by potentiometric titration. The benefits resulting from this technique are explained in section 1.5.

1.3. Buffer capacity

Wines are buffer solutions because they contain a mixture of weak organic acids and conjugated bases (Chapter 1.1), which limits (buffers) the change in pH following addition of a base or an acid. The buffer capacity is a measure of the resistance a wine or similar solution exerts against pH changes. It is defined as the amount of OH^- ions necessary to raise to pH by 1 unit, or the amount of H^+ ions required to lower the pH by 1 unit. Thus, the units of buffer capacity are moles H^+ ions (or OH^- ions) per liter per pH unit (M/L/pH). However, because of the actual values of buffer capacity in juices and wines, they are usually expressed in millimolar (mM) terms, and these are generally in the range of 30 to 50 mM/L/pH unit.

In summary, high buffer capacity values predict a smaller change of pH following a given change of TA. It is important to note that buffer capacity is positively correlated with total acidity, i.e. the sum of free and neutralized acids. It has some importance in winemaking when acid adjustments are performed: Wines or musts with larger buffer capacities require greater additions of acid to lower pH.

The analytical determination of buffer capacity is easy to perform: A juice or wine sample is simply titrated in either direction for a full pH unit and the volume of titrant noted. This can be done directly to the sample without caring about dissolved CO_2. However, buffer capacity is not regularly measured in the realm of wine quality control. In common practice, it is accepted as it is because other goals take precedence. Therefore, it is difficult to exactly predict the pH changes caused by acidification or deacidification procedures.

1.4. Individual acids, their dissociation and strength

Wine contains six acids in sufficient amounts to play a role in the intensity of perceived sourness. Two among them, tartaric and malic acid, are already present in all grapes and musts. Their concentrations start to decrease after the onset of primary fermentation. During alcoholic fermentation, further acids such as citric, succinic, and acetic acid are produced by yeasts.

Malic acid ($COOH$-$CHOH$-CH_2-$COOH$) is a diprotic (divalent) acid, because it carries two functional carboxyl (-$COOH$) groups. It is present in grapes at high concentrations exceeding 20 g/L prior to veraison, but is actively metabolized in grapes and declines during ripening. Therefore, concentrations are generally in the low range of 1 to 4 g/L in musts obtained from ripe fruit in the warmer growing areas, whilst they can exceed 8 g/L in musts from underripe fruit in cool-climate regions. Some yeast strains are able to partially metabolize malic acid during alcoholic fermentation. During malolactic fermentation, it can be completely converted into lactic acid (Chapter 10).

Tartaric acid ($COOH$-$CHOH$-$CHOH$-$COOH$) is also a diprotic acid and characteristic of grapes but is not found in other common fruits. It is poorly metabolized during grape ripening and largely constant on a per berry basis, but declines with ripening due to berry growth and dilution. It is not metabolized during winemaking at pH levels below 4.0, but can suffer heavy losses through precipitation of its insoluble salts. In freshly pressed juices, it occurs at levels of between 3 to 8 g/L. For sound grapes, malic and tartaric acid dominate TA before fermentation, compromising more than 90 % of it.

Citric acid ($COOH$-CH_2-$COH(COOH)$-CH_2-$COOH$) is a triprotic acid present in quite low concentrations in young wines. Its concentration in musts rarely exceeds 0.5 g/L and is further lowered during malolactic fermentation.

Lactic acid ($COOH$-$CHOH$-CH_3) is monoprotic and essentially the result of the transformation of malic acid by lactic acid bacteria during malolactic fermentation by bacteria and can reach levels of up to 4 g/L, depending on the initial malic acid concentration. Only a few 100 mg/L can be produced during alcoholic fermentation.

Succinic acid ($COOH$-CH_2-CH_2-$COOH$) is diprotic and produced by yeasts during alcoholic fermentation in a range normally varying from 0.5 to 1.0 g/L, though some yeast strains are able to produce up to 3 g/L under certain fermentation conditions.

Acetic acid ($COOH$-CH_3) is normally produced up to 0.1–0.5 g/L by yeasts during alcoholic fermentation, but its concentration can increase as a consequence of bacterial development. Thus, Acetobacter bacteria produce it in conjunction with ethyl acetate via oxidation of ethanol under aerobic conditions, whilst some strains of malolactic bacteria can also produce it through degradation of residual sugar in the case of sluggish or stuck fermentations. It is the major constituent of volatile acidity (VA). In high concentrations, it is an index of low quality in wines, so that VA is subject to legal restrictions, in general 1.08 g/L (as acetic acid) for white and 1.20 g/L for red wines.

In wines made from rotten fruit, there is a more complex array of acids. When bacterial infection occurs in damaged grapes, the resultant juice can initially contain up to 3 g/L of acetic acid. In wines produced from fruit affected by noble rot, concentrations of several g/L have been measured for gluconic acid (Amann 2007), of up to 0.4 g/L for mucic acid (Clauss et al. 1966), and of up to 0.6 g/L for citric acid.

In contrast, minor levels of volatile fermentation-derived acids and phenolic acids occur in the mg/L concentration range and do not significantly affect titratable acidity. Free sulfurous acid (SO_2) and ascorbic acid can only be found in wine after they have been added as enological supplements. Their current concentrations in wine do not have a significant impact on TA (Chapter 1.1). Amino acids have a similar low impact because of their incorporation into the yeast cell mass during fermentation and their low strength. The negative charge of their carboxyl groups is partially compensated by the positive charges of their amino groups.

pK_A values indicate the strength of acids

As previously stated, the various acids in wine display different strengths, which result from their different readiness to ionize and dissociate H^+ ions. The dissociation, i.e. the extent to which acids release H^+ ions, depends on their chemical structure and the initial pH of the solution. The lower the pH or the more H^+ ions are already present in the solution, the less the relatively weak organic acids are ready to dissociate and donate their H^+ ions.

The ability of an acid to dissociate is indicated by its pK_A value (acid dissociation constant). The pK_A is the pH at which half of the acid is ionized, that is, at which its acid group has donated 50 % of its H^+ ions able to dissociate.

For any pH, the dissociated percentage of a weak acid can be calculated by the Henderson-Hasselbach equation dealt with in standard chemistry textbooks. From this equation, it follows that

- when pH is equal to pK_A, 50 % of the acid is dissociated,

- when pH is 1 unit lower than pK_A, 10 % of the acid is dissociated,

- when pH is 1 unit higher than pK_A, 90 % of the acid is dissociated.

Polyvalent acids as tartaric or citric acid have more than one acid group, which dissociate to a different extent. Hence, they are characterized by various pK_A-values, according to the number of acid groups they contain. The lower the pK_A, the stronger the acid is or its respective acid group. Table 2 shows the pK_A-values of the most important acids found in wine as well as the names of their anions and salts.

acid	1st acid group (pK$_{A1}$)	2nd acid group (pK$_{A2}$)	3rd acid group (pK$_{A3}$)
tartaric acid (tartrates)	3.04	4.34	
malic acid (malates)	3.48	5.10	
lactic acid (lactates)	3.86		
citric acid (citrates)	3.13	4.76	6.39
acetic acid (acetates)	4.76		
gluconic acid (gluconates)	3.86		
succinic acid (succinates)	4.16	5.61	
ascorbic acid (ascorbates)	4.25	(11.57)	
phosphoric acid (phosphates)	2.16	7.21	(12.32)
sulfurous acid (sulfites)	1.81	7.20	
sulfuric acid (sulfates)	~ 1	1.06	
carbonic acid (carbonates)	6.35	(10.33)	

Table 2: The strength of different acids in wine expressed as their pK$_A$-values in water at 20° C, and designation of their salts.

Remark: pK$_A$ values of acid groups in brackets are completely neutralized in the pH-range of wine and are not included in titratable acidity (TA) measurements.

Tartaric acid will be used to give an example of variable dissociation and its dependence on pH. When tartaric acid dissociates the H$^+$ ion of its first acid group, the remaining anion is called "bitartrate" or "hydrogen tartrate". When it splits off the H$^+$ ions of both its acid groups, the remaining anion is called "tartrate". Hence:

tartaric acid (TH$_2$)		bitartrate ion (TH$^-$)		tartrate ion (T^{2-})
COOH | HCOH | HCOH | COOH	$\xleftrightarrow{\text{pK}_A = 3.04}$	COO$^-$ + H$^+$ | HCOH | HCOH | COOH	$\xleftrightarrow{\text{pK}_A = 4.34}$	COO$^-$ | HCOH | HCOH | COO$^-$ + H$^+$

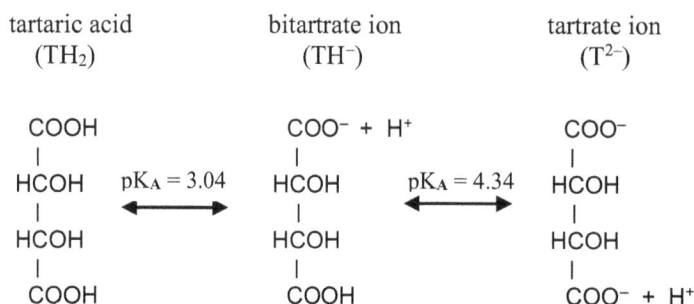

Using once more the example of tartaric acid, figure 4 shows the relative concentrations of it undissociated and its ionized forms as a function of pH. This graph can be found in most enological textbooks. It is based on the pK$_A$ values of tartaric acid in aqueous solution given in table 2.

Figure 4: Relative concentrations of undissociated tartaric acid (TH$_2$), bitartrate anion (TH$^-$), and tartrate anion (T^{2-}) in aqueous solution.

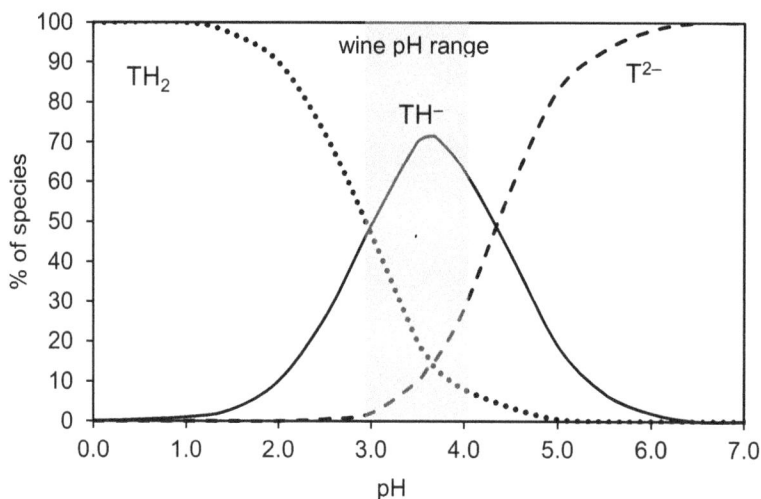

The first dissociation constant (3.04) of tartaric acid falls within the normal range of wine pH (3.0 – 4.0), and its second dissociation constant (4.34) in a pH range slightly higher than that found in wines. This means that tartaric acid at wine pH exists mainly in its half dissociated (TH$^-$) bitartrate form, whilst its undissociated (TH$_2$) and completely dissociated (T^{2-}) forms make up a considerably lower percentage.

Effect of alcohol content and ionic strength on pK$_A$ values

It must be emphasized that the pK$_A$ data reported in table 2 refer to water at 20° C used as a reference solvent. They can be found, with minor variations, in almost all enological textbooks, neglecting the fact they change in the presence of the significant alcohol and sugar levels in wines and musts. These changes are due to changes in dielectric (insulating) properties of the solvent and its effect on the strength of bonds within the carboxyl groups of the acids.

Experimental determination of the effects of ethanol concentrations on the pK$_A$ values has been made for the major organic acids of wine (Usseglio-Tomasset & Bosia 1978). The results have attracted far less attention in the wine industry than they deserve. They are reported in table 3.

Table 3: The strength of different acids expressed as their pK$_A$ values in ethanol solutions (Usseglio-Tomasset and Bosia 1978).				
Acid	12 % ethanol		14 % ethanol	
	pK$_{A1}$	pK$_{A2}$	pK$_{A1}$	pK$_{A2}$
tartaric acid	3.23	4.59	3.27	4.63
malic acid	3.64	5.32	3.67	5,36
citric acid	3.32	4.91	3.35	4.95
succinic acid	4.38	5.84	4.41	5.90
lactic acid	4.06		4.09	
acetic	4.89		4.91	
gluconic acid	3.99		4.03	

The effect of ethanol in wine is to make the effective pK$_A$ values 0.15 to 0.2 units higher than their aqueous values. This leads to less dissociation of the acids than can be observed in water solution.

Besides alcohol, the so-called ion activity or ionic strength also has an impact on the pK$_A$'s of acids. This term is to be seen against the background that ions do not show an ideal, linear behavior corresponding to their concentration. In contrast, they lose a part of their activity with increasing concentration. This deviation from the ideal behavior can be taken into account by using activity coefficients. However, many texts that describe acidity calculations do not take them into account.

The ionic strength effect on acid dissociation is especially pronounced for the first pK$_A$ of the acids, and this leads to much higher pH values for the maximum bitartrate ion concentration, often referred to as pH 3.6 or 3.7 based on aqueous pK_A's, but more like 4.1 when ion strength and ethanol effects are taken into account (Boulton et al. 1996). Under these conditions, the curve for bitartrate (TH^-) shown in figure 4 shifts to the right over higher pH values, yielding figure 5.

Figure 5: Relative concentration of the bitartrate anion (TH⁻) of tartaric acid taking into account average wine alcohol content and ionic strength.

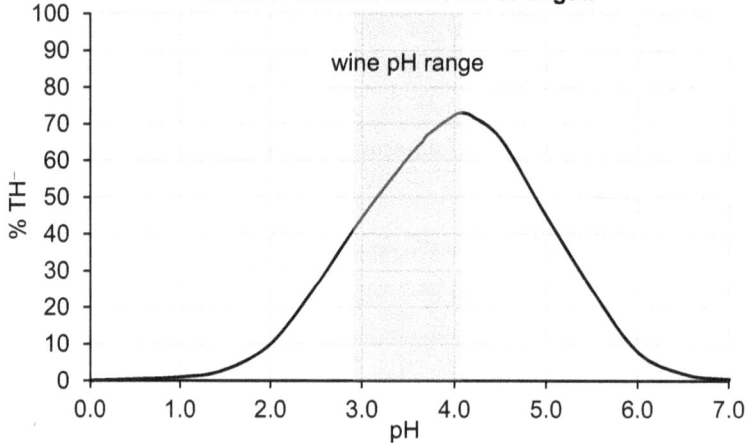

Practical importance of tartaric acid dissociation

The pK_A data for tartaric acid have considerable practical significance. This applies in particular to the pH shift observed when potassium bitartrate precipitates spontaneously or via crystal stabilization procedures (Chapter 11.2).

The pK_A values reported in tables 2 and 3 also show that in the pH range of wines, tartaric acid is significantly more dissociated than malic, citric, or lactic acid. Consequently, a TA with a high percentage of tartaric acid leads to a lower pH than a TA with a low portion of tartaric acid. This explains why acidification of must and wine is preferably performed using tartaric acid when the intention is to decrease the pH, thereby improving microbial stability (Chapter 3.5).

Another feature of tartaric acid with far-reaching enological importance is that it is the only acid in wines obtained from sound fruit able to lead to the precipitation of poorly soluble salts. These salts are potassium bitartrate (Chapter 11), calcium tartrate (Chapter 12) and calcium tartrate malate (Chapter 8.2). Their concentrations depend on the extent to which tartaric acid is ionized. For that reason, the bitartrate and tartrate percentages of total tartaric acid as shown in figures 4 and 5 are important key figures in the evaluation of crystal stability.

The low solubility of the tartaric acid salts also forms the basis on which chemical deacidifications are carried out. Their basic feature is a removal of tartaric acid by precipitating it as salts when potassium or calcium ions are added (Chapter 6).

Another acid able to sometimes produce insoluble salts is mucic acid, a bivalent C6-acid. It is only produced by the *Botrytis cinerea* fungi grown on rotten grapes, probably by enzymatic oxidation of galacturonic acid split off from pectin. When mucic acid content is more than 0.1 g/L, it can drop out as calcium mucate, its barely soluble calcium salt (Clauss et al. 1966). Deposits in some Botrytis-style wines are characterized by a white, amorphous powder rather than a defined crystal shape.

1.5. The mineral cations and their impact on pH and acidity

By nature, wines contain four mineral cations of more or less importance according to the individual wine:

Sodium (Na^+) most frequently occurs at concentrations below 100 mg/L (Berg et al. 1979), but in wines made from grapes grown using irrigation with high salinity water, in coastal areas, or sprayed with sodium carbonates used in organic viticulture, contents can largely exceed 100 mg/L. Legal limits vary widely among countries.

Magnesium (Mg^{++}) is to be found at concentrations ranging from 50 to 100 mg/L. No significant viticultural or enological impacts on its content are currently known.

Calcium (Ca^{++}) occurs by nature at concentrations of 60 to 120 mg/L. Deacidification procedures using calcium carbonate can give rise to significantly higher contents (Chapter 6.4). Calcium uptake by filtration or fining with bentonites rarely exceeds 20 mg/L (Klenk and Maurer 1969), but can rise to more than 40 mg/L by fining with particular bentonite brands. In a few cases, elevated calcium concentrations of more than 200 mg/L have been observed and could not be explained by calcium uptake during winemaking. The impact of soil is poorly investigated.

Potassium (K^+) is the major cation in ripe grapes, accounting for about 70 % of the total mineral cations. In freshly pressed musts obtained from unmacerated grapes, it varies in a broad range of concentration from 400 to 2,000 mg/L, depending on a large number of viticultural variables. Its level is further differentiated by grape processing techniques (Chapter 2.3). Subsequently, concentrations in wine are strongly influenced by enological treatments such as acidification (Chapter 3.4), chemical deacidification (Chapters 6.3 and 7.1), and bitartrate stabilization (Chapter 11).

Influence of potassium on pH and TA

During ripening, the vines take up potassium - and to a lesser extent sodium and other cations - from the soil and store it in the grapes. As a consequence, H^+ ions are replaced by potassium ions carrying likewise a positive charge by means of a biochemical exchange mechanism across cell membranes leading to expulsion of H^+ ions from the grape (Boulton 1980 c). The removal of H^+ ions leads to an increase of pH. Concurrently, TA decreases since it only measures acids whose dissociable H^+ ions have not been exchanged for potassium ions. In other words, a part of titratable acidity is neutralized by potassium. The relationship between "titratable acidity" (free acids) and "neutralized acids" as depicted in figure 1 shifts towards neutralized acids.

Without that partial exchange of protons supplied by the acids for potassium and other cations, the pH would correspond to that of the pure acid mixture, about pH 2.2 for the acid mixture of wines. In contrast, a complete exchange would lead to complete neutralization and a pH of approximately 8. The pH range of 3.0 o 4.0 represents a partially neutralized acid mixture, in which the extent of exchange is between 20 and 30 %.

The pH of wine results essentially from the interaction between potassium and acids. In years with high amounts of precipitation in late summer, water mobilizes soil potassium and promotes its uptake in the vines. Under these conditions, a high pH may occur despite high TA figures. This is a frequent problem in cool-humid wine growing areas. In contrast, wines obtained under cool-dry growing conditions tend to display less potassium and correspondingly lower pH data. On the other hand, wines grown under hot-dry conditions are also expected to have less TA and, consequently, higher pH values. All in all, potassium affects pH as much as does TA, although in the opposite direction (Boulton 1980 b, Kodur 2011).

Generally, the inverse relationship between pH and TA is not as close as expected by many practitioners. As shown in figure 6, in an assessment of 55 bottled wines from Europe, only 80 % of the pH could be explained by TA data. This is the most important reason why pH and TA should be reported simultaneously and interpreted in their mutual context.

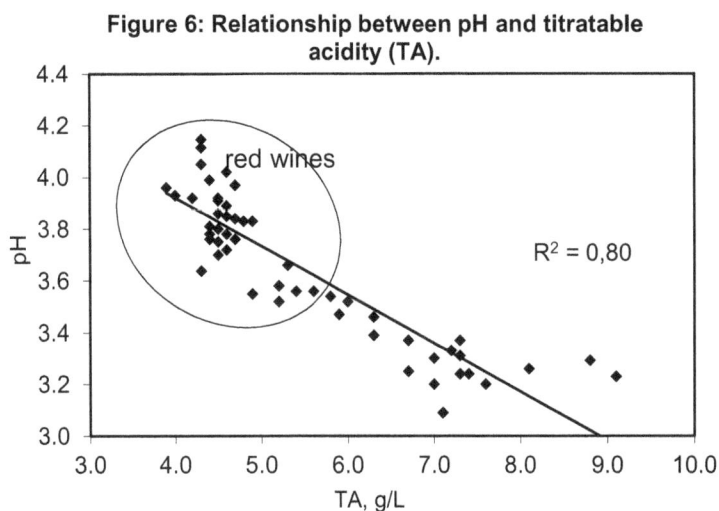

Figure 6: Relationship between pH and titratable acidity (TA).

The pH figures of most wines vary between 3.0 and 4.0. Outliers of higher than pH 4.0 can occasionally be found in red wines, when a low TA coincides with high potassium content. Outliers downwards (pH < 3.0) occur sometimes in white wines and are indicative of a strong acidification with tartaric acid (Chapters 3.4 and 3.5), but can also be ascribed to excessively high crop yields or unripe fruit.

Other cations such as calcium, magnesium, and sodium act in a way comparable to that of potassium. However, their effect is far less pronounced since their concentration in the grapes is less important and less dependent on soil humidity.

The potassium-acidity balance

Storage of mineral cations such as potassium in the grapes equates to deacidification happening in the vineyard. Taking into account the balance between acids and mineral cations, there are different ways of acidity correction in must and wine:

- An acidification can be achieved by addition of an acid that remains stable in the wine. It can also be performed by a partial removal of mineral cations, for example by precipitation of potassium with tartaric acid (Chapter 3.4) or by binding them to a cation exchange resin (Chapter 4.1). Thereby, originally neutralized acids are released from their salts. The relationship between free acids and neutralized acids shifts towards free acids without changing their overall concentration expressed as total acidity (Figure 1).

- The same principle applies to chemical deacidifications. Its most current implementation relies on a removal of free acids by their precipitation as insoluble salts after adding cations such as potassium or calcium. However, the precipitation intended to occur may fail partially or completely. In such a case, a portion of the acids is just neutralized; the relationship between free acids and neutralized acids shifts towards neutralized acids without changing their overall concentration (Figure 1).

2.

Sensory fundamentals of acidity management

Frequently, measures to correct acidity focus on specific target values for pH and total acidity. Sensory features, which alone determine the market value of wine, are neglected in this technocratic approach. The following chapter deals mainly with sensory issues such as the sensory properties of the various acids and mineral cations and their interaction. Special attention is paid to the sensory consequences of variable potassium contents and how these are influenced by measures of acidity management. Examples of simple sensory exercises are given that allow understanding how these effects are perceived on the palate.

2.1. Sensory properties of the acids

Tartaric, malic, lactic, and acetic acid are the most important acids in wine from a quantitative point of view. Most of their designations derive from the Latin names of the fruit or foodstuff they have first been identified in. This connection frequently boosts assumptions about their sensory properties being reminiscent in some way of their origin. While associations of that kind can still be found in some opinion-leading standard readings, upon wine tasting, stringent sensory evaluation unravels any rumors and esoteric assumptions: Lactic acid is not mild or reminiscent of milk, citric acid does not evoke any citrus flavor or fresh taste, and malic acid does not taste "green" or convey the smell of green apples.

Malic acid has an especially bad reputation in the minds of some opinion leaders. The assumption that malic acid is harsh and that tartaric acid tastes ripe is one of the most persistently circulating rumors in some sectors of the wine industry. In reality, both acids are not distinguishable one from another when tasted at an equal concentration expressed as titratable acidity. This means that it is not of sensory importance whether 1 g/L of TA consists of tartaric or an equivalent amount of malic acid (von Nida and Fischer 1999 a). This also means that the bad reputation of malic acid is not caused by any special taste properties, but just by high amounts occurring in musts and wines made from unripe fruit.

All acids mentioned previously simply taste only more or less sour, and most but not all of them do not display any gustatory or olfactory side activities related to their origin. However, they differ as to their strength from both a sensory and a chemical point of view. These differences cannot be explained only by their number of acid groups or their pK_A values (Tables 2 and 3) because their respective molecular weights also matter. Therefore, conversion of one acid into another is only possible on the basis of their equivalent weight. The equivalent weight of an acid is calculated as the quotient of its molecular weight : number of acid groups.

For easy conversion of one acid into another, the respective conversion factors are reported in table 4.

Table 4: Interconversion of acidity units expressed in terms of one acid to another.

Known expression (acid)	Conversion factor for expressing acidity as								
	tartaric acid	malic acid	citric acid	lactic acid	acetic acid	sulfuric acid	succinic acid	gluconic acid	ascorbic acid*
tartaric	-	0.893	0.853	1.200	0.800	0.653	0.787	2.614	2.358
malic	1.119	-	0.955	1.344	0.896	0.731	0.881	2.972	2.106
citric	1.172	1.047	-	1.407	0.938	0.766	0.922	3.060	2.012
lactic	0.833	0.744	0.711	-	0.667	0.544	0.656	2.175	2.830
acetic	1.250	1.116	1.066	1.500	-	0.817	0.984	3.264	1.887
sulfuric	1.530	1.366	1.306	1.837	1.225	-	1.204	3.995	1.540
succinic	1.271	1.135	1.084	1.525	1.017	0.830	-	3.319	1.856
gluconic	0.383	0.342	0.327	0.460	0.306	0.250	0.301	-	1.856
ascorbic	0.424	0.379	0.362	0.509	0.339	0.277	0.334	1.108	-

Example: TA of a wine is 6.5 g/L (as tartaric acid). Expressed as citric acid it would be (6.5 x 0.853 = 5.5 g/L (rounded).

*Determined by TA titration after addition to wine.

The different strengths of the acids can easily be understood by tasting them in aqueous solutions. In these solutions, 1 g/L of lactic acid tastes less sour than 1 g/L of the stronger malic acid. Arranging the randomly presented samples in order of increasing sourness will be easy a task. However, this finding is only half of the truth.

Comparing sourness of acids requires a common basis

The basic question is another one. Is it possible to distinguish the various acids when they are all adjusted to a comparable concentration of 1.0 g/L titratable acidity by dissolving 1.00 g/L of tartaric acid, 0.893 g/L of malic acid, 1.20 g/L of lactic acid, or 0.853 g/L of citric acid?

The large number of studies (Pangborn 1963, Sowalsky and Noble 1998, Co-Seteng et al. 1989, Rubico and Daniel 1992) published on this issue provides quite different results. The most realistic answer is that when even trained tasters compare these solutions side by side in one flight, they are not able to distinguish whether the sour taste elicited by them is caused by one or another acid. Until now and despite the unmanageable number of trials conducted with blind tastings, nobody in the wine industry has been able to detect a difference between for example 0.89 g/l malic acid and 1.20 g/l lactic acid in water, let alone in wine. In contrast, all winemakers and enologists involved in real-world winemaking and participating in these trials only marvel at the manifold, fanciful and often contradictory descriptors assigned to the individual acids in the technical literature and even in articles with scientific pretensions, all in connection with wine. They range from "soft, mild, slightly sweet, spicy, smooth with persistent sourness, with a mild dairy aroma" for lactic acid, "harsh, metallic, mellow, persistent sourness reminiscent of green apples" for malic acid, "hard, mineral, fresh, citrus-like, dry, brusque sourness that dissipates quickly" for tartaric acid to "clean, bright, refreshing, with pleasant citrus note" for citric acid. The suspicion arises that one copies the other's nonsense without doing any precise research, as is so often the case in wine-specific literature.

In summary, when the intention is to reproduce a comparable sour taste, tartaric, malic, lactic, citric, and even phosphoric acid can be arbitrarily replaced one by another as long as the conversion factors reported in table 4 are taken into account. On a molar concentration basis and in the concentration range present in wine, they just taste equally sour, as was correctly stated in at least one (Johanningsmeier et al. 2005) of the papers on this topic.

Sensory exercise 1 allows reproduction of the result outlined above.

Sensory exercise 1: Comparison between different acids in aqueous solution at 1.0 g/L titratable acidity (as tartaric acid).

- 1.000 g/L tartaric acid

- 0.893 g/L malic acid

- 0.853 g/L citric acid

- 1.200 g/L lactic acid

- 0.800 g/L acetic acid

However, there are also exceptions which must be taken into account, because some acids can actually display secondary sensory qualities going beyond their basic taste of sourness. This property is well known from acetic acid. Since it is volatile, it has also an olfactory activity. Furthermore, it evokes a pungent, scratching sensation in the aftertaste after oral ingestion. Dependent on the type of wine, these sensory side qualities of acetic acid arise at concentrations of more than 0.4 to 0.7 g/L. It is useless to try to give an exact threshold, because in any case it depends strongly on the wine matrix.

Additionally, succinic acid is known for its bitter aftertaste (Rubico and McDaniel 1992). Its low concentration of 0.3 to 1.5 g/L in typical wine does not contribute significantly to any perceived bitterness (Amann 2007). Some bitterness in wine occasionally ascribed to gluconic or any other acid is not attributable to these acids, but rather to high potassium, tannin, or alcohol levels.

In this context, particular attention must be given to the outstanding behavior of tartaric acid. It is the only acid occurring in wine able to lower potassium levels by precipitation. After adding tartaric acid to a wine, the sensory effects resulting from potassium depletion are in no way related to any specific taste properties of tartaric acid.

When a wine is supplemented with a standard amount of 150 mg/L of ascorbic acid, titratable acidity is increased by 0.063 g/L (as tartaric acid). Ascorbic acid is sometimes added to white wines as a reducing agent and oxygen radical scavenger in order to improve their shelf life. The amounts added only marginally contribute to sour taste to an extent that most often is not significant.

The causes of the sour taste

The sour taste of acids has various causes and is by no way only related only to the H^+ ions they donate. When the acids occurring in wine are dissolved to prepare solutions of comparable normality concentrations, for example of n/50, and neutralized with potassium or sodium ions to pH 7.0, acetic acid loses any sour taste, while the other organic acids continue to display a mild sour taste. This behavior shows that the sour taste of acetic acid is exclusively caused by its undissociated molecules, while the sourness of the other organic acids is at least partially elicited by their anions.

A further trial confirms the ability of undissociated acid molecules to evoke a sour taste: An n/1000 solution of hydrochloric acid, completely dissociated and displaying a pH of 3.0, hardly appears sour. On the other hand, solutions of the weaker tartaric or acetic acid at identical pH but at necessarily higher concentrations exhibit a markedly sour taste whose intensity increases when the concentration of the undissociated acid molecules is enhanced (Peynaud 1987).

The present state of knowledge allows us to hypothesize that the intensity of perceived sourness of acids depends on their concentration, their chemical structure, their strength, and on the concentration of H^+ ions (pH) of the solution. In no case, can the sour taste be deduced from a single one of these parameters

(Pangborn 1963). On the contrary, it is essentially characterized by the interaction of titratable acidity and pH (Amerine et al. 1965, CoSeteng et al. 1989).

For sour taste in real wines, titratable acidity (TA) is considered much more important than pH (Plane et al. 1980). In a sensory study comprising 46 wines from various countries, TA accounted for 77 % of the sensory perception of sourness (von Nida and Fischer 1999 b). This result is substantiated by two further, simple observations: In wines with a given TA but with a pH considered extremely low or high for that TA, the intensity of sourness continues to be governed more by TA than by pH. Furthermore, when a wine is diluted with water, pH remains almost constant, while TA and sour taste decrease by the dilution factor (Boulton 1980 b).

To take into account of the impact of pH on wine sourness, a so-called "'acidity index" has been developed. According to this index, the sensorially perceived sourness correlates with the difference of "TA – pH" (Plane et al. 1980).

There a several explanations of why TA is so important for perceived sourness. One of them is its partial titration that occurs in the mouth by the saliva, which is slightly basic due to the bicarbonate anions it contains. The sensation of the presence of the wine causes a saliva flow somewhat in proportion to the quantity of neutralization required. This is generally correlated with the TA. Variable saliva flow rates between individuals explain why there is considerable variation in the perception of sourness between judges, similar to the perception of astringency (Schneider and Tracey 2021).

By comparison, the pH is merely an indication of the extent to which the acid mixture of wine has been neutralized by mineral cations taken up by the grape or provided by acidity adjustments. It is not correlated with the amount of acids present, but is more influenced by their ability to dissociate protons, i.e. their strength.

Astringency caused by acids

Acids do not only elicit a sour taste, but also evoke a sensation of astringency. This sensation is not to be confounded with any of the primary tastes (sweet, sour, bitter, salty, and umami), but represents a tactile perception of dryness, roughness, and contraction on the mucous membranes of the mouth and throat. Its feeling is best experienced upon ingestion of tannins in red wines, where it goes together with a simultaneously generated bitterness. In the context of sensory training, the most suitable way of modeling pure astringency is the use of solutions of aluminum potassium sulfate, which does not elicit any bitter primary taste.

Some acids are even perceived as more astringent than sour. In such a case, the relationship between sourness and astringency depends markedly on the acid concentration. In acid mixtures, astringency correlates with the concentration by $r = 0.55$. On the other hand, the astringency elicited by those solutions is more strongly affected by pH than is their sourness (Rubico and McDaniel 1992).

While the perception of sourness is simultaneously influenced by the type of acid, its concentration, and pH, astringency evoked by acids is almost exclusively dependent on pH. It increases when pH decreases (Sowalsky and Noble 1998). This strong dependence on pH proves that the astringency of organic acids stems directly from their acidic properties. It is not the result of a chemical reaction of the acids with saliva proteins, which is responsible for the astringent effect of tannins (Lawless et al. 1996).

2.2. Sensory properties of the mineral cations

Sodium and magnesium at concentrations commonly occurring in wines (Chapter 1.5) are without any particular sensory importance. At best, they slightly contribute to neutralize acids. Recognition thresholds of sodium chloride (NaCl) responsible for perceived saltiness have been reported to be 2.05 g/L in white and 1.77 g/L in red wine, respectively (de Loryn et al. 2014). Such elevated NaCl contents may occur in wines as a result of high soil or water salinity when irrigation is practiced. Hence, these thresholds do not refer to any specific taste properties of sodium at common concentrations of less than 100 mg/L Na^+.

It is important to remember that it is illegal to add sodium based additives to wine.

Calcium can become sensorially relevant by its intrinsic taste properties. In wine sensory sciences, its taste is only marginally discussed. It can be described as mealy, floury, sticky, gluey, grinding, scratching, and reminiscent of diatomaceous earth. Frequently it is confused with the astringency of tannins, the pungent and scratching aftertaste of volatile acidity, or the burning sensation elicited by elevated alcohol contents. The perception of this very specific mouthfeel can easily be trained by tasting wines spiked with increasing amounts of calcium ions and stabilized with metatartaric acid for a short time. Details of this procedure are given in sensory exercise 2.

Sensory exercise 2: Taste of increasing concentrations of calcium in a standard dry white wine (< 80 mg/L Ca^{++}) with constant titratable acidity.

Calcium is incorporated into wine by dissolving $CaCO_3$ and neutralized by adding equivalent amounts of malic acid in order to keep TA constant. Samples should be prepared within one or two days before tasting and stabilized temporarily by addition of metatartaric acid (200 mg/L) to prevent uncontrolled calcium precipitation.

– white wine
+ 0 mg/L Ca^{++}

– white wine
+ 100 mg/L Ca^{++} (+ 0.25 g/L CaCO$_3$, + 0.336 g/L malic acid)

– white wine
+ 200 mg/L Ca^{++} (+ 0.50 g/L CaCO$_3$, + 0.672 g/L malic acid)

– white wine
+ 300 mg/L Ca^{++} (+ 0.75 g/L CaCO$_3$, + 1.007 g/L malic acid)

Trained tasters are able to detect calcium contents of approximately 200 mg/L Ca^{++}. Concentrations above 250 mg/L Ca^{++} are perceived as objectionable in the aftertaste in almost any kind of wine. The calcium detection threshold is subject to the rules of adaptation. This means that after repeated ingestion of high calcium wines, tasters get used to it.

Generally, the typical taste of calcium only appears when abnormal high concentrations occur after deacidification with calcium carbonate (Chapter 8.4).

Potassium is of large sensory importance owing to its intrinsic taste, its concentration as well as to the extent it neutralizes acids. It substantially shapes wine style on the palate and is considerably affected by means of acidity management. Therefore, its role deserves a more detailed review given in section 2.3.

No link between mineral cations and perceived minerality

As a matter of principle, the mineral cations are by no means related to the notion of minerality. Since these cations are not volatile and therefore without any smell, they cannot contribute to the olfactory perception of what is called minerality. However, this is a term that is widely used in contemporary wine description and promotion, albeit a kind of buzzword without any clear meaning. Its inflationary use tends to summarize a broad range of olfactory, retronasal, and gustative perceptions without duly distinguishing between all these sensations. Sometimes it is also misused to euphemistically qualify reduction taints. Hence, it is an ill-defined term calling for a linguistic adjustment and exact definition.

There is a wide consensus that the perception of minerality on the palate is enhanced by a high titratable acidity (Deneulin et al. 2014, Heymann et al. 2014). In contrast to popular belief, it is not related to the taste of minerals transported through the vine from the vineyard rocks and soils and stored in the grapes, because these minerals represent complex crystalline structures of high molecular size that cannot be taken up and transported by the roots. Although attempts to explain the perception of minerality involve allusions to geological materials, these are irrelevant to its origin from a geological perspective (Maltman 2013). The minerals contained in wine such as potassium, sodium, calcium etc. are alkali or alkaline metallic cations not related to geological minerals. Furthermore, they are not involved in the taste sensation of what is described as salty.

Although potassium and calcium carry a specific taste, their concentrations do not correlate with minerality or saltiness perceived on the palate. From the chemical category of cations and acids, sensory studies only were able to confirm the role of a high titratable acidity in the perception of minerality on the palate. So far, acidity management can influence perceived minerality.

In olfactory terms, minerality seems related to reductive taints and a low scoring of fruity descriptors (Ballester et al. 2013). It is also often associated with attributes like wet stones, wet sand, conglomerate pebbles, chalk, iodine, oyster shells, flint stones struck against themselves, struck matches, smoke, and high levels of free SO_2. Studies on chemical substances eliciting specific aroma attributes described as mineral are scarce and limited to two compounds - benzene methanethiol and hydrogen disulfane. Details are given in Schneider (2019).

Since the minerality term lacks a precise semantical definition, it is hardly suitable as an accurate wine flavor description despite its popular use. All studies underline the role of the tasters' cultural background in the definition of perceived minerality and the underlying wine composition, thus explaining divergent results. The only sure fact in this context is that the mineral cations contained in wine do not contribute to it.

2.3. Enological differentiation and sensory qualities of potassium

Potassium is a sensory component that has been underestimated for many years, but there is an emerging appreciation of its role in the gustative perception of wine.

As outlined before, titratable acidity is the most reliable parameter to describe the intensity of sour taste. Potassium is one of the wine constituents relativizing the absolute validity of this rule. Besides its role in the neutralization of acids, potassium and its salts also result in a particular taste, which is able to partially mask the sour taste of TA (Schneider 1998). Its content is affected by technical means, both acidification and deacidification measures, and the stage at which they are performed. Consequently, the final TA achieved by adjustments can taste more or less sour. Acidity management and potassium management are closely associated one with another.

The taste of potassium

The sensory properties of potassium on the palate can be studied by spiking a wine with increasing amounts of potassium without modifying its TA. For that purpose, potassium is added in the form of one of its neutral salts, for example potassium malate. Alternatively, potassium carbonate or bicarbonate can be added in conjunction with an equivalent amount of malic acid. Under these conditions of comparable TA, the difference threshold of potassium is approximately 200 mg/L in both white and red wines (Schneider 1998). This kind of experimental approach mimics the same conditions occurring in wines with naturally elevated potassium contents. They are held in solution by malic acid or any other soluble acid.

From a sensory point of view, high levels of potassium salts increase the perception of gustatory parameters such as body, weight, volume, and fatness. They contribute much more to these sensory perceptions than the legal additives like mannoproteins, tannins, or gum arabic usually recommended for that purpose. In some way, they are also comparable to those elicited by high (~ 20 g/L) glycerol contents. Last but not least, they are strongly affected by the kind of acid used for acidification (Chapter 3.4) or the time of chemical deacidification (Chapter 6.4). Whenever acidity management is deemed necessary, it can be used to exert a lasting influence on potassium salt concentration.

It is important to note that when potassium exceeds a certain concentration, the wines are described as soapy, bitter, salty, fat, or oily, while the untreated standard is considered as more filigree, crispy, delicate, subtle but also more meager. Implementation of sensory exercise 3 gives an introduction into this matter

---

Sensory exercise 3: Impact of increasing concentrations of potassium (K^+) on the in-mouth sensations of a standard white wine (< 700 mg/L K^+) at identical titratable acidity.

Potassium is added as potassium bicarbonate ($KHCO_3$) and neutralized by additions of equivalent amounts of malic acid.

- white wine
  + 0 mg/L K^+
- white wine
  + 300 mg/L K^+ (+ 0.77 g/L $KHCO_3$, + 0.517 g/L malic acid)
- white wine
  + 600 mg/L K^+ (+ 1.54 g/L $KHCO_3$, + 1.034 g/L malic acid)
- white wine
  + 900 mg/L K^+ (+ 2.31 g/L $KHCO_3$, + 1.551 g/L malic acid)
- white wine
  + 1200 mg/L K^+ (+ 3.08 g/L $KHCO_3$, + 2.068 g/L malic acid)

Note that in this exercise, 1.0 g/L $KHCO_3$ can be replaced by 0.69 g/L K_2CO_3 (potassium carbonate). The exercise should be repeated on a red wine.

Apparently, every single wine shows a sort of optimal potassium content, at which it is considered typical by hedonic conventions. In a broader sense, when the focus is on the production of varietal wines, an optimum potassium content can be ascribed to each variety. Clearly, potassium as such is an underestimated gustatory element in enology, and this is not only because it is related to pH.

Similar trials revealed that increasing potassium content by 500 mg/L at identical TA lowers the intensity of perceived sourness to an extent that corresponds to 0.5 g/L TA (Schneider 1998). This amount of potassium is removed from solution when 1.9 g/L tartaric acid or 0.95 g/L TA precipitate as potassium bitartrate, for example in the course of crystal stabilization. On the other hand, a surplus of 500 mg/L potassium remains in solution when 1.9 g/L tartaric acid is removed by must deacidification. These observations explain why a wine with a given TA may taste more or less sour according to its potassium content.

Differentiation of potassium by viticultural variables

Potassium uptake by the roots and its accumulation in grapes and must is affected by cultivar, rootstock, the size and age of the root system, the number of clusters on the vine, canopy management, soil, soil moisture and precipitations, and the year of harvest. Among these factors, rootstock and the availability of soil moisture are the most crucial ones. The selection of low potassium-accumulating rootstocks is seen as part of the strategy to overcome high pH effects, because high pH figures are predominantly attributable to high potassium concentrations, as will be illustrated in figure 8. Appropriate rootstocks have been identified (Kodur 2011, Walker and Blackmore 2012).

High precipitation rates, particularly during the ripening period, or vineyard irrigation boost potassium contents. In contrast, untreated wines from grapes harvested from dry soils or harvested in dry years are characterized by considerably lower levels of potassium (Mpelasoka et al. 2003). Heavy, moist soils lead to higher potassium uptake and storage in the grapes than light or stony soils.

Differentiation of potassium during vinification

The transfer of grape potassium into must and wine is influenced by winemaking procedures. Free-run juices contain significantly less potassium than the respective pressings. Figure 7 gives an example. Increasing mechanical load on the must during consecutive pressings and pressure increases fosters the extraction of potassium that, in turn, enhances pH and lowers TA to an extent which is not negligible. In general, the last press fraction displays at least 1 g/L TA less than the free-run juice.

Figure 7: Impact of fractions of a membrane press on potassium concentration and pH of a Riesling must as compared to the free-run fraction.

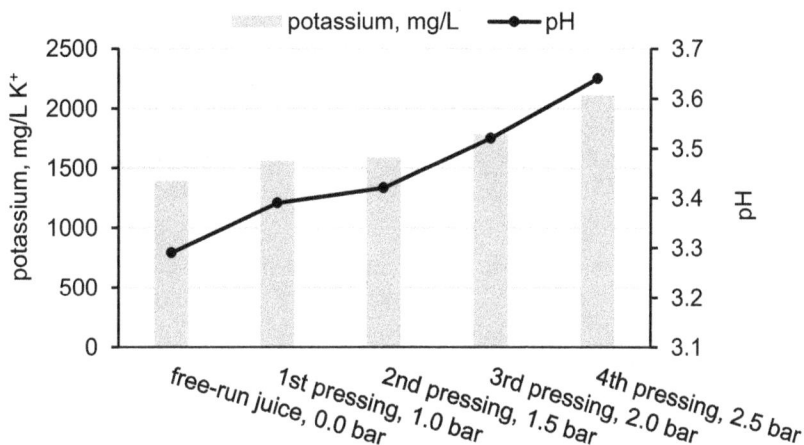

Skin contact for several hours increases potassium levels in the wines by 5 to 20 % via extraction from the skins. This extraction is combined with a decrease of titratable acidity in the range of 1 to 2 g/L (von Nida and Fischer 1999 b), which is to be explained by neutralization and partial precipitation of tartaric acid.

Long skin contact time occurring in red wine making, as well as mechanical treatments of must by stirring and pumping, leads to a further increase of potassium levels by 50 to 100 % during the first two or three days of fermentation, after which the potassium concentration plateaus or starts do decline (Harbertson and Harwood 2009). Therefore, red wines tend to display higher potassium levels in conjunction with higher pH values than white wines. Figure 8 illustrates this relationship by a set of randomly selected bottled Central European wines. In this set of samples, 79 % of pH is explained by potassium content.

Figure 8: Potassium content of white and red wines and its relationship with pH.

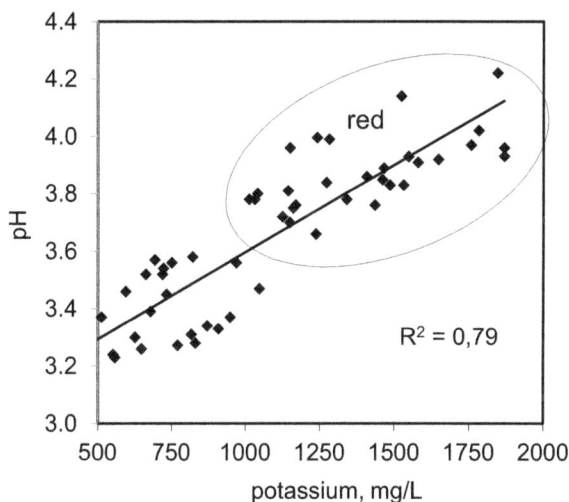

The multitude of influencing factors makes it easy to understand why the natural potassium content of freshly pressed juices varies within a large range between 1,000 and 2,500 mg/L. Stoichiometrically, this corresponds to 1.9 to 4.8 g/L of acidity (as tartaric acid) that becomes neutralized. Compared with this, the other mineral cations remain in the background.

Effect of potassium bitartrate precipitation on potassium and TA

During and after alcoholic fermentation, considerable amounts of potassium precipitate in conjunction with tartaric acid as potassium bitartrate (potassium hydrogen tartrate). With each g/L of tartaric acid dropping out, 262 mg/L of potassium are removed.

Potassium hydrogen tartrate is an acidic salt, in which only one of the two acid groups of tartaric acid is neutralized. Hence, when 1 g/L of tartaric acid is removed from the system, TA (as tartaric acid) decreases by 0.5 g/L. This results in the following quantitative relationships for potassium bitartrate precipitation:

- 1.0 g/L tartaric acid precipitates with 262 mg/L potassium to 1.262 g/L potassium bitartrate, resulting in a loss of 0.5 g/L titratable acidity.

- A 1.0 g/L decrease of TA caused by potassium bitartrate precipitation involves a loss of 2.0 g/L of tartaric acid and of 524 mg/L of potassium.

- The precipitation of 1.0 g/L of potassium bitartrate reduces tartaric acid by 0.8 g/L, titratable acidity by 0.4 g/L, and potassium by 207 mg/L, respectively.

During spontaneous potassium bitartrate precipitation after alcoholic fermentation of white wines, a decrease of titratable acidity by 1.0 to 2.5 g/L can usually be observed. These losses are reinforced when targeted measures of crystal stabilization are taken to remove potassium bitartrate present in surplus (Chapter 11.2). In red wines, losses of titratable acidity by potassium bitartrate precipitation tend to be less since their tannin-anthocyanin aggregates hamper crystallization.

An example for clarification: Analysis of a white grape juice obtained under cool-climate conditions yields 6.0 g/L tartaric acid and 11.0 g/L TA. After spontaneous potassium bitartrate precipitation in the young wine, titratable acidity has become reduced by 2.0 g/L, leaving only 9.0 g/L TA. The difference indicates that tartaric acid has been reduced by $2.0 \cdot 2 = 4.0$ g/L and potassium by $4 \cdot 262 = 1,048$ mg/L K^+. Therefore 2.0 g/L of tartaric acid remains. By definition, this example is a simplified approach. In reality, losses of titratable acidity by potassium bitartrate crystallization are partially overlapped by the synthesis or degradation of organic acids by yeast metabolism during alcoholic fermentation (Chapter 5.2).

Using a mathematical model, pH and titratable acidity of wine has been predicted based on juice data for pH, titratable acidity, tartaric acid, and malic acid. Juice potassium was derived from pH and the individual organic acids. Average deviation between predicted data and measured data in wine was 0.1 for pH and 1.0 g/L for titratable acidity (Höchli 1997). Such a mathematical model promotes the understanding of what is happening to the anion-cation balance when must turns into wine and the quantitative relationships involved, but it is not really useful under practical winemaking conditions.

The losses of tartaric acid and TA observed during the spontaneous crystallization of potassium bitartrate in young white wines depend essentially on the initial concentration of potassium available for precipitating tartaric acid. This is the reason why some varieties or some wines with low initial potassium levels keep more tartaric acid in solution than others. As a general rule it can be stated that young white wines contain approximately half of the tartaric acid and potassium contents of the respective juices. However, this estimate is not valid when juice is subjected to measures of acidification or deacidification.

Effect of potassium bitartrate precipitation on pH

The effect of potassium bitartrate (KHT) precipitation on pH is variable, depending on how much of this salt is removed and on the initial pH.

As can be deduced from figure 5 and table 3, the maximum bitartrate concentration for a given level of tartaric acid happens to be at the midpoint between the pK_A values of the first and the second acid group of tartaric acid, i.e. around pH 4.1. Thus, when KHT precipitates at an initial pH lower than 4.1, which is the case in almost any wine, pH decreases. The reason is that for every molecule of KHT that precipitates, one free H^+ ion originally attached to the tartrate in KHT is released. In contrast, when KHT precipitates in one of the rare wines with pH above 4.1, the pH increases even more as one H^+ ion is incorporated into KHT and removed from solution.

These figures and ratios are in contrast to the statement, which was often made in the enological literature in the past and still is today, according to which the maximum bitartrate concentration occurs at pH 3.65. In accordance with this outdated theory, the pH value would only decrease if the initial pH were below 3.65, and it would increase if it were above 3.65. This pH is called the tipping point of KHT.

However, if the pK_A values of tartaric acid are corrected for alcohol and ionic strength of wine according to table 5, the tipping point or maximum bitartrate concentration shifts to pH 4.1. Observations under commercial winemaking conditions confirm that the latter value is the correct one: In practically all wines, the precipitation of KHT leads to a decrease in pH. This decrease is greater the lower the initial pH. It diminishes with increasing pH until at pH 4.1, where the pH reduction due to KHT precipitation is practically zero. This is of particular importance when tartaric acid is added to lower the pH (Chapter 3.4).

The magnitude of pH shift also depends on the amount of KHT that drops out, irrespective of whether this happens spontaneously during fermentation or induced by cold stabilization. It is not correlated with the decrease of perceived sourness observed when KHT precipitates. This decrease better correlates with the loss of TA associated with KHT precipitation.

Impact of acidity adjustments on potassium levels

Cool-climate growing conditions frequently require a chemical deacidification of juice or wine, especially white wines. When it is carried out on juice, a large fraction of tartaric acid is removed. Obviously, this fraction is no longer available to generate KHT crystals and decrease potassium post fermentation. Therefore, after severe deacidification of juice no further decrease of titratable acidity by KHT precipitation can be observed in the young wine. Concurrently, potassium is fixed at its initial high level. In contrast, when chemical deacidification is postponed until after fermentation, potassium can become lowered in the meantime by spontaneous KHT precipitation.

It follows from the above that when chemical deacidification is deemed necessary, the timing of its implementation allows for substantially shaping the final potassium content of wine. This fact explains one of the basic differences between juice and wine deacidification by chemical means (Chapter 4.4). Regardless of final TA, the choice between juice and wine deacidification has sensory consequences, which are attributable to the content and specific taste of potassium and its soluble salts. High potassium juices benefit from wine deacidification, while juice acidification is the better solution when one aims at preserving potassium. These options are of great importance in cool-climate areas and in grape varieties that often require chemical deacidification.

In contrast, hot-climate growing conditions often suggest acidification of musts. It is most frequently carried out by adding tartaric acid to the must since this acid is most suitable to achieve low pH, recommended for microbial stability. However, most of the tartaric acid added to musts drops out as KHT, removing equivalent amounts of potassium. In contrast, other acids do not affect potassium concentration. Hence, the kind of acid used for acidification is an essential tool for potassium management (Chapter 3.4) no matter if the acid addition is performed to must or wine.

Treatment of high potassium wines

When high potassium concentrations exert a negative effect on the palate, the most common treatment is an addition of tartaric acid to partially precipitate it as KHT. As long as there is no knowledge about potassium concentration, additions of about 2 g/L tartaric acid would be a good guess. Details are discussed in Chapter 3.4. Cation exchange can be a viable alternative (Chapter 4.1). Unavoidably, both kinds of treatment also increase titratable acidity, which might create a problem in high potassium - high TA wines. Bipolar electrodialysis (Chapter 4.2) allows for decreasing potassium concentration without significantly increasing TA.

3.

Acidification by supplementation with organic acids

Acidification has always been an important issue in hot-climate regions, and in times of climate change it is also gaining importance in cool-climate regions. Addition of organic acid is the most common means of correcting low TA and high pH figures, and various acids are available for this purpose. Their technical properties, enological implications, and sensory consequences are discussed in this chapter. Special emphasis is placed on the frequently employed approach of acidification with tartaric acid from the perspective of pH decrease, increase of titratable acidity, and losses of potassium with the sensory consequences associated therewith. In this chapter the ever-debated question of safety through a pH considered low enough is also critically evaluated, without attaching to it the exaggerated relevance it has in wide circles of the wine industry and enological teaching. Instead, the focus is on sensory analysis, and numerous suggestions are made for performing and evaluating preliminary bench trials for sensory optimization, without overemphasizing analytical data in a simplistic way.

3.1. Legal aspects

Premature harvest can be a measure for preserving natural grape and must acidity levels particularly in hot growing areas. On the other hand, it is also associated with the risk of vegetal-green aromatics, harsh tannins in red wines, or atypical aging in white wines. Furthermore, global climate changes since the early 1990's has led to higher average temperatures and extreme climatic events in many wine growing areas of the world. As a result, low TA and high pH values are increasingly occurring, thereby requiring remedial must or wine acidification even in cool-climate growing areas historically associated with high acidity levels. Thus, acidification has turned into a global challenge, though it does not necessarily mean dumping bins of acid into the must just to achieve a pH thought to be safe.

The USA, Canada and New Zealand allow the addition of tartaric, malic, lactic, citric, and fumaric acid.

In Australia and South Africa, tartaric, malic, lactic, and citric acid can be added.

In the EU, tartaric, malic, and lactic acid are approved in amounts not exceeding 1.5 g/L (as tartaric acid) when added to must and 2.5 g/L when added to wine, making up a total of 4.0 g/L. The American regulation, instead, allows acidification on musts and wines up to a maximum level of fixed acidity in the finished wine of 9 g/L, expressed as tartaric acid.

According to European (OIV) and European law, citric acid is not considered just an acidulant, but a stabilizer. In Europe it is allowed for 'wine stabilization purposes' due to its ability to chelate and stabilize certain metal ions such as iron up to a maximum residual level of citric acid in the final wines of 1.0 g/L. Within this limit, its acidic action is used as a fortunate side effect.

The addition of mineral acids such as hydrochloric or sulfuric acid is not allowed anywhere.

Calcium sulfate, the neutral calcium salt of sulfuric acid, lowers the pH in wine without increasing the titratable acidity. Therefore, its use is not considered as a means of acidification. It has been traditionally used in the production of sherry in Spain and has limited applicability. Details are given in Chapter 3.5.

3.2. Technical properties of the acids

Some of the acids approved for acidification differ in their optical forms from those found in nature. For instance, there are L-malic acid and D-malic acid. The prefixes D and L specify the optically active form of acids that can have more than one form. Such acids are optical isomers (enantiomers) because they possess at least one asymmetrically substituted carbon. They distinguish themselves by their plane of polarization of linearly polarized light, which they turn either to the left or to the right. However, their chemical, physical, and sensory properties are identical.

Tartaric acid

Tartaric acid is approved for acidification only as L-tartaric acid. It is obtained from grapes and byproducts of winemaking. It provides the strongest pH decrease. Therefore, it is preferred for treatment of musts displaying a high pH that is expected to create microbial risks during alcoholic and malolactic fermentations. It is also of interest for acidification of very flat wines with potassium levels high enough to be disturbing on the palate. It is the only acid whose addition leads to a significant precipitation of potassium bitartrate, accompanied by an equivalent decrease of potassium, tartaric acid, titratable acidity, and extract (Chapter 2.3). This behavior results in sensory side effects and technical consequences, which distinguish tartaric acid from other acids and require a more detailed discussion (Chapter 3.4). Tartaric acid is not subject to microbial degradation, at least not under hygienically sound winemaking conditions.

Malic acid

Malic acid in commercial amounts is only available as DL-malic acid. This designation refers to a so-called racemic mixture, in which the two optically active isomers of malic acid occur in identical amounts. In contrast, malic acid occurring in nature and grapes only includes the L-malic acid. Only the latter can be broken down into lactic acid during malolactic fermentation, while D-malic acid is not subject to microbial degradation. Owing to its microbial instability, the use of DL-malic acid for acidification is only recommended after filtration or addition of sulfur dioxide to the wines. It is soluble and cannot drop out as crystals.

Lactic acid

Lactic acid also shows two optical isomers, both occurring in wine. L-lactic acid is produced by bacterial degradation of L-malic acid during regular malolactic fermentation, but the generation of D-lactic acid requires degradation of glucose by bacteria. Levels of more than 0.5 g/L of D-malic acid provide evidence for serious microbiological spoilage and are usually accompanied by high levels of volatile acidity. For acidification, L-lactic acid is used.

In contrast to tartaric, malic, or citric acid, lactic acid does not exist in a crystalline form, but is commercialized as an 80% solution. Therefore, its use requires a conversion factor of 1.25 which is roughly canceled out by its higher density of 1.2 g/mL. All things considered, one mL of commercial lactic acid of 80 % contains 0.96 grams pure lactic acid, which amounts to 0.80 grams titratable acidity expressed in tartaric acid.

Lactic acid is esterified with itself to an extent of 7 to 10 %. This ester splits within a couple of days after addition to the wine. After that, the complete acidification effect appears. Sensorially, the effect is not really significant.

Furthermore, lactic acid is easy to handle, microbiologically stable, and does not precipitate. Frequently, it displays a lactic odor, which cannot be detected in the wine after its addition. Concerns that lactic acid causes a lactic off-flavor reminding one of a badly conducted malolactic fermentation (MLF) are not justified. Lactic acid is a pure acid, while the buttery, lactic flavor occurring in some wines after MLF is to be ascribed to diacetyl and can be an undesirable side product of MLF.

Citric acid

Since citric acid is degraded during MLF, it is used most frequently in wines after SO_2 addition and filtration. It is soluble and does not produce any crystals. Commercial citric acid is a monohydrate containing one water molecule per molecule of citric acid so that its actual content is only 91.4 % of pure citric acid.

3.3. Implementation of sensory trials for wine acidification

Tartaric acid is preferably used for must acidification because it best contributes to microbiological stability owing to the relatively strong decrease in pH it causes (Section 3.5). However, after filtration and sterile bottling, this effect loses importance. In the almost finished wine shortly before bottling, the option of minor sensory improvements on the palate comes to the foreground. The use of acidification for that purpose is a powerful but widely underestimated tool of sensory optimization. In contrast to acidification of musts, acidification of wine has the advantage of allowing to precisely adjust perceived sourness, balance, and other in-mouth sensations. This approach is often based on modest additions of acids other than tartaric acid, does not necessarily rely on numbers, and requires careful optimization by sensory trials.

Malic, lactic, citric, and fumaric acid remain stable in solution and do not drop out as does tartaric acid. The final TA adjusted to remains stable. Therefore, these acids are most suitable for minor acidity corrections upwards in the vast majority of wines where such an effect is deemed desirable. Exceptions apply when distracting high potassium levels are to be lowered simultaneously (Chapters 2.3 and 3.4).

The optimal acidification range is difficult to calculate from analytical figures. In every single wine, the addition of X g/L of an acid or the adjustment to Y g/L final TA leads to a different sensory outcome. The differences arise from variable matrix effects, which mask perceived sourness to a variable extent. In wines obtained under hot-climate growing conditions, 1 g/L TA tends to be sensorially more dominant than in wines grown under cool-climate conditions. Acidification only relying on figures (pH, TA) frequently results in wines presenting a sharp sourness on the palate. Preliminary tests are highly recommended.

Tests of this kind suggest the prior preparation of a 10 % aqueous stock solution of the acid chosen for the purpose, for instance 100 g/L of malic or citric acid or 125 mL/L of lactic acid 80 %. Since differences in the strength of acids are less important in this context aiming exclusively at achieving a defined sensory objective, malic, lactic, and citric acid can be replaced one by another. Addition of 1 mL of such an acid solution to 100 mL of sample equals an addition of 1 g/L of the respective acid to the wine. Adjustment of various acidity levels can be performed according to sensory exercise 4 on the following page.

Sensory exercise 4: Preparation of test solution for acidification of wine and implementation of trials.					
Preparation of test solution	Dissolve 100 g of citric acid* and adjust to 1000 mL with distilled water.				
Use of the test solution	**0.1 mL / 100 mL wine equals in the tank:** + 0.1 g/L citric acid				
	+ 0.1 mL	+ 0.2 mL	+ 0.3 mL	+ 0.4 mL	+ 0.5 mL
Equals in the tank:	+ 0.1 g/L	+ 0.2 g/L	+ 0.3 g/L	+ 0.4 g/L	+ 0.5 g/L
*Remark: Citric acid can be replaced by equivalent amounts of malic or lactic acid according to table 3.					

During this procedure, one will find out that

– acid additions of more than 1 g/L often results in a taste markedly too sour and sharp;

– additions in the range of 0.2 to 0.5 g/L result in a balanced sourness and taste in numerous wines;

– additions of only 0.1 g/L lead to a distinguishable increase of perceived sourness;

– the increase of perceived sourness is not linear to the increase of TA or the decrease of pH.

Increasing the acid addition sensorially judged optimal by approximately 10-20 % is an acceptable and reasonable measure. This is explained by the fact that after some weeks or months, the amount of acid added becomes better integrated in the wine and loses its initial tartness. Esterification in conjunction with a very slight decrease of TA is apparently involved in this effect (Edwards et al 1985).

Every winemaker should have such an acid solution prepared and at hand. Its use helps the winemaker create a more intimate relationship with his wines, understand why they differ one from another, and learn more about the daily fluctuations in his sensory perception.

For the fine tuning of an almost finished wine, the use of tartaric acid can be useful in individual cases if, in addition to acidification, a reduction of sensorially disturbing high potassium contents is intended at the same time. The sensory effects are covered in the next section. Whenever tartaric acid is used, the subsequent precipitation of potassium bitartrate and loss of TA must be taken into account. Therefore, test samples and untreated control should be frozen overnight to accelerate the precipitation of unstable bitartrate and the achievement of the final TA. Before tasting and after thawing, bitartrate crystals must be removed immediately by decanting or filtration. Otherwise they would redissolve and enhance TA quite fast when temperature rises. The reasons are explained in the following section.

3.4. Specific features of acidification with tartaric acid

Superficially, acidification with tartaric acid looks like an easy procedure to master. On the surface, it appears that the addition of 1.0 g/L of tartaric acid increases TA likewise by 1.0 g/L, and it actually does so for a short time. However, the long term effect will be different. As outlined previously (Chapter 2.3) and in contrast to other acids, the special feature of tartaric acid is that it produces insoluble salts and drops out. This happens in conjunction with potassium. The salt that precipitates is potassium bitartrate (KHT). Consequently, potassium content also decreases.

KHT has acid character since only one of the two acid groups of the bivalent tartaric acid is neutralized by potassium, while the other one continues to act as an acid. Thus, each g/L of tartaric acid that drops out leads to a loss of 0.5 g/L of titratable acidity and 262 mg/L of potassium (Chapter 2.3).

The aforementioned behavior explains why the addition of 1.0 g/L of tartaric acid only temporarily increases TA by 1.0 g/L. The subsequent precipitation of KHT inevitably leads to a partial loss of the gain in TA. The extent of that loss depends on how much of the tartaric acid added will actually drop out as KHT. Two extreme situations can emerge:

– The tartaric acid added completely drops out as KHT. In such a case, addition of 1.0 g/L of tartaric acid results in a persistent increase of TA by only 0.5 g/L.

– The tartaric acid added entirely remains in the wine because it is prevented from precipitation, for example by the presence of metatartaric acid, carboxymethylcellulose (CMC), potassium polyaspartate, or any other colloids able to inhibit crystallization (Section 11.3). Under these conditions, addition of 1.0 g/L of tartaric acid generates a permanent increase of TA by 1.0 g/L.

In practice, the TA increase achieved varies between these two extreme values. It depends on the concentration of potassium that is available for precipitation and the technical conditions affecting that precipitation.

Under common winery conditions, one may simplify and assume that the addition of 1.0 g/L of tartaric acid increases TA by approximately 0.6 g/L on average. In unfiltered wines and at current cellar temperatures (10 to 15° C) this effect only arises after some weeks of further storage. Deliberate cold stabilization shortens this period by accelerating KHT precipitation. Hence, after tartaric acid additions, KHT precipitation rate and timing of sampling greatly influence TA and pH figures. Therefore, when running sensory trials using tartaric acid according to sensory exercise 4, precipitation has to be accelerated by severe cooling or freezing before sensory evaluation of the test samples takes place.

Acidification with tartaric acid is not only an option for optimizing perceived sourness. The decrease of potassium content it generates as a side-effect has far reaching consequences, which stem from the intrinsic taste properties of potassium.

Losses of potassium generate more vibrant, filigree, and crispy wines

Increasing potassium concentration without altering TA results in an increase of gustatory perceptions described as body, volume, and weight on the palate accompanied by a decrease of perceived sourness (Chapter 2.3 and Sensory exercise 3). Decreasing potassium content generates wines characterized as leaner, more filigree, vibrant, and crispy, but also more meager.

Since acidification with tartaric acid usually lowers potassium concentration, it comprises much more than just an increase of sourness. The removal of associated potassium lowers the sensation of volume on the palate, but also that of soapiness in wines when excessively high potassium levels are detrimental to quality. Sensory exercise 5 helps us understand how acidification with tartaric acid differs from that with other acids due to the removal of potassium it causes.

Sensory exercise 5: Comparison between tartaric and citric acid for acidification of wine.			
Preparation of test solutions	Dissolve 100 grams of citric acid* and adjust to 1,000 mL with distilled water. Dissolve 100 grams of tartaric acid and adjust to 1,000 mL with distilled water.		
Treatment of wine	Reference, untreated white wine **	+ 0.69 g/L citric acid	+ 1.50 g/L tartaric acid. Freeze overnight, thaw, and decant.
Increase of TA (in g/L tartaric acid) ***	0	0.75	0.75
Decrease of potassium (in mg/L K$^+$) ***	0	0	393
Remarks: * Citric acid can be replaced by an equivalent amount of malic or lactic acid. ** The wine should be filtered, contain no crystallization inhibitors (Chapter 11.3), and display pH > 3.6 and TA < 5.5 g/L. *** Calculated figures presuming that tartaric acid added precipitates totally			

Figure 9 depicts the results of a sensory study carried out on a Chardonnay whose initial TA of 6.0 g/l was adjusted to 6.5 g/L by adding either 0.45 g/L malic acid or 0.90 g/L tartaric acid plus cold stabilization. For the same final TA achieved, sensory differences were caused by differences in potassium concentration, which was reduced by 235 mg/L. Final pH was slightly higher after acidification with malic acid than with tartaric acid (pH 3.48 vs. 3.42). This example clearly shows that the obsessive search for low pH values is not compatible with the commitment to producing full-bodied wines.

Figure 9: Sensory impact of acidification of a dry Chardonnay from 6.0 g/L to 6.5 g/L TA using a) 0.9 g/L tartaric acid and cold stabilization, and b) 0.45 g/L malic acid.

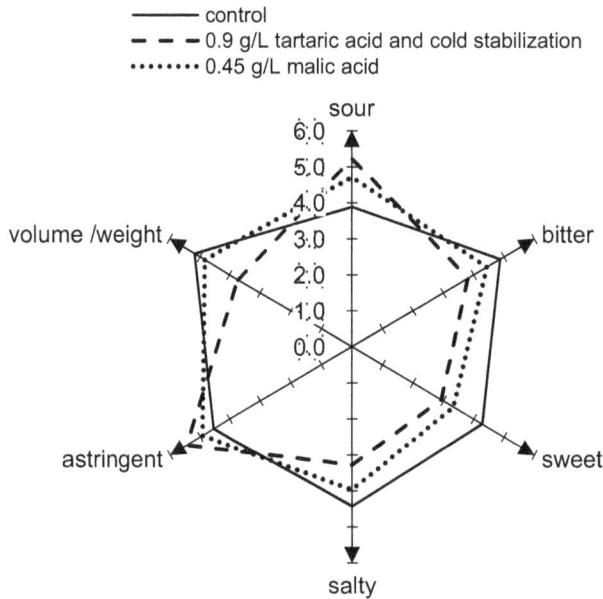

On a merely sensory level, the effect of tartaric acid additions is more distinctive in wines with naturally low potassium levels as frequently obtained under hot-climate growing conditions. In those wines, acidity tends to come more to the fore because masking effects exerted by potassium are canceled. Conversely, acidification with tartaric acid will be of interest for wines displaying high potassium levels considered objectionable due to their soapy aftertaste.

In conclusion, it can be summarized that acidification using tartaric acid is only useful when

– high potassium contents sensorially deemed detrimental are to be lowered;

– high pH figures allow for anticipating microbiological complications, particularly in musts or when unfiltered wines with low free SO₂ levels are to be aged in barrels.

In all other cases, the use of lactic, malic, or citric acid makes more sense, in particular when working on wines. It simplifies preliminary trials and has only a minor effect on crystal stability. When acidifying white musts, many wineries have achieved good sensory results with a combination of tartaric acid and another soluble acid because the potassium, and thus the volume on the palate, was not reduced as much as when tartaric acid was used alone.

Measuring and correctly interpreting potassium concentration units is a good way to make the right choice between tartaric and any other acid used for acidification. In practice, this knowledge generally does not exist and must be replaced by another approach:

Figure 10: The relationship between potassium content and the ratio 'pH : TA'.

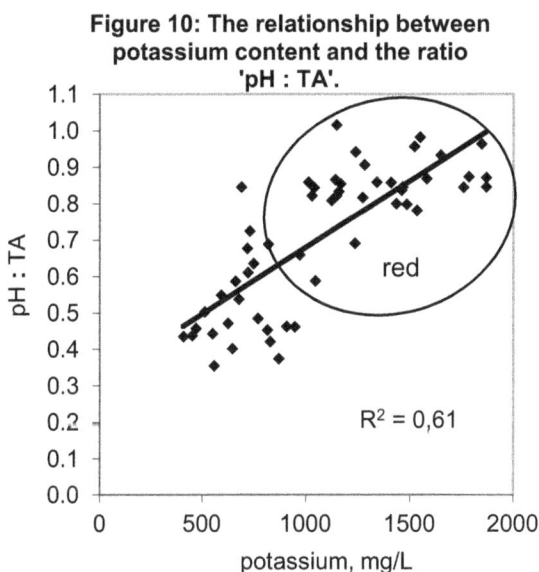

The basic analytical feature of high potassium wines is a relatively high pH for a given TA. Figure 10 shows this relationship measured on bottled wines of Central European origin. In this set of samples, 61 % of the potassium content is explained by the ratio of pH to TA. In other words: If the pH appears suspiciously high in relation to the titratable acidity, the suspicion of a high potassium content is justified.

Remedying high potassium/ high TA wines

When elevated potassium concentration is actually perceived as detrimental depends on masking effects exerted by the wine matrix. Hence, no fixed potassium thresholds can be given, but concentrations of more than some 1,700 mg/L are considered to be high.

Sometimes, excessively high potassium levels can be found in wines obtained from juices that have undergone chemical deacidification prior to fermentation. The reason is that early withdrawal of tartaric acid stabilizes potassium at its high must level (Chapter 6.4). The most frequent reason, however, is a humid season, which leaves enough malic acid able to keep potassium in solution. These conditions are particularly feared in some cool-climate growing areas with heavy soils.

For most high potassium wines, the most approachable way of lowering potassium is the addition of tartaric acid as soon as possible. Tartaric acid precipitates potassium the better the more one approaches pH 4.1, the peak of the bitartrate curve (Figure 5). Frequently, the increase of TA caused by that approach is affordable. But what happens when excessively high potassium concentrations coincide with high TA?

On one hand, these high pH/high TA wines require deacidification to adjust TA to a desirable level. On the other hand, they also suggest addition of tartaric acid to lower potassium levels and prevent excessive soapiness on the palate. Tartaric acid additions increase TA even more. This kind of enological dilemma cannot be overcome by the use of a traditional cation exchange resin exchanging potassium for hydrogen ions since it would also result in a TA increase (Chapter 4.1).

In most wineries, the only means to resolve this problem would be a tartaric acid addition followed by a deacidification that, after the depletion of tartaric acid, is able to remove malic acid. A double-salt deacidification carried out in the so-called extended version (Chapter 6.3) is one way to achieve that purpose. It's a kind of treatment that strips out lots of aromatics when one does not master gentle wine processing.

Another way is biological deacidification (Chapter 10). In some wineries, malolactic fermentation (MLF) is still associated with huge impacts on wine style and aroma alterations when carried out on white wines. However, progress in selecting appropriate bacteria strains has been enormous so that more and more winemakers feel comfortable sending their white wines through MLF. The only important thing is that the decision about MLF must be made in a timely manner before any SO_2 additions.

In the meantime, technical progress is opening up new prospects for treatment of high pH/high TA wines. The emerging techniques are all based on physical means. One of them involves anion exchange with the resin in the tartrate form followed by sending part of this wine through a cation exchange with the resin in the H^+ form and a final blending the two portions. The tartrate anion exchange reduces the pH of the wine by replacing malic acid with tartaric acid. Precipitation of the majority of the added tartrate as potassium bitartrate decreases the TA of the wine. However, nonspecific removal of flavor and color compounds by the resins has negated the practical use of this technique (Bonorden et al. 1986).

Bipolar electrodialysis is another, even more sophisticated technique to simultaneously adjust any cations and anions, pH included. It is briefly presented in chapters 4.2 and 9.2.

Timing of acidification

In hot-climate growing areas, tartaric acid additions to musts are a standard operation. They have a twofold purpose: They lower pH at the earliest possible moment to improve microbial stability during a microbiologically critical stage of production, and they compensate for deficiencies in TA and sourness expected to occur in the future wine.

While the first objective is relatively easy to achieve, adjusting to a harmoniously balanced sourness is more difficult to achieve by early tartaric acid additions based only on pH and TA figures. It is an almost impossible endeavor that is successful only by chance. As tartaric acid precipitates for the most part but not to a calculable extent, resulting losses of TA, potassium, and perceived sourness in the finished wine are hardly predictable. Therefore, when lowering pH or potassium levels is not a major concern during vinification, sensory fine tuning of the almost finished product using acidification with malic, lactic, or citric acid according to sensory exercise 5 would be the better solution.

Sensory evaluation after bitartrate stabilization (Chapter 11) and shortly before bottling provides a more secure basis for acidity fine tuning. When doing so, a sensory comparison with reference wines, for instance from the previous year or bottling, provides more valuable help than merely relying on analytical figures. All will depend on the expectations producers have on the quality of their wines.

Sometimes, it is useful to perform both a must and a wine acidification in a cumulative way. Otherwise, there is a risk of obtaining over-acidified and badly balanced wines when too much tartaric acid is added to the must. In such a case, a slight deacidification with potassium carbonates can provide a remedy.

3.5. Impact of acidification on pH

When an acidification is carried out with the purpose of lowering pH, the question is how much of a given acid has to be added in order to decrease pH to a desired extent. The answer is not unequivocal and depends on the buffer capacity of each wine.

Buffer capacity is by definition the mole equivalents of H^+ or OH^- ions that are required to change 1 L of solution by 1 pH unit. It is expressed in mmol/pH. In other terms, it describes the resistance to changes in pH upon addition of an acid or a base (Chapter 1.3).

Buffer capacity of wines depends essentially on TA and its makeup by individual organic acids, while potassium and other minerals content exert less influence. The more acid a wine contains, the more changes in pH are buffered or compensated. When in the course of TA titration a wine is deprived of H^+ ions by neutralization with the OH^- ions of the base added, pH first increases and causes the organic acids to release more H^+ ions that make pH drop again. Therefore, the titration curve ascends so slowly in the range of pK_A values of tartaric and malic acid (Figure 2).

Only when the buffer capacity is known we can compute how much acid has to be added to adjust, for instance, a must from pH 3.8 to pH 3.3. Or, the other way around, by how many g/L TA a wine has to be deacidified to enhance a very low pH to pH 3.2 necessary to start malolactic fermentation (Chapter 10.1). The buffer capacity can be calculated from analytical data or measured directly; respective procedures are given in the standard chemistry literature. However, under practical winemaking conditions, empirical data are used:

Addition of 1 g/L titratable acidity in the form of equivalent amounts of malic or lactic acid lowers pH by approximately 0.1 unit. However, when 1 g/L titratable acidity is added as 1.0 g/L of tartaric acid, pH is decreased by 0.15 to 0.20 units. This is related to the fact that the first acid group of tartaric acid shows the lowest pK_A value of all these acids and releases the most H^+ ions in the pH range of wine. This effect is all the stronger the higher the initial pH.

The major portion of tartaric acid added drops out with potassium as potassium bitartrate. This leads to a further decrease in pH as long as the pH before precipitation does not exceed 4.1 (Chapter 2.3). From a sensory perspective, these pH changes are less important than the losses of TA caused by bitartrate precipitation.

Because of the almost entire precipitation of added tartaric acid in musts and most wines, such an addition is difficult to prove by analytical means.

The special case of gypsum

The addition of calcium sulfate ($CaSO_4 \cdot 2 H_2O$, gypsum), also called "plastering", is a very ancient and particular technique of acidification. It is still legalized for the treatment of musts intended for the production of Sherry and other dessert wines in Spain, provided that the residual sulfate content does not exceed 2.5 g/L expressed as potassium sulfate. It is commonly added in conjunction with tartaric acid.

Gypsum lowers the pH without making any change to titratable acidity. This effect is based on the conversion of potassium bitartrate (KHT), a weakly acid salt, into equivalent amounts of tartaric acid (TH_2), with potassium sulfate (K_2SO_4) and calcium tartrate (CaT) generated as by-products according to the overall formula

$$CaSO_4 + 2 KHT \rightarrow TH_2 + K_2SO_4 + CaT_\downarrow$$

(Gomez Benitez et al. 1993). In the broadest sense, it could also be said that sulfuric acid added as its calcium salt partially replaces the weaker tartaric acid.

The exchange of one mole of tartaric acid for one mole of sulfuric acid plus the slight increase of free tartaric acid explain the decrease of pH, which is approximately 0.12 units per 1 g/L calcium sulfate added, compared to 0.17 units per 1 g/L of tartaric acid. The combined addition of gypsum and tartaric acid produces an additive effect when compared to the doses of each one individually. This allows a reduction in the doses of tartaric acid necessary to achieve a pH of microbiological interest, whilst more potassium is kept in solution instead of precipitating as potassium bitartrate. There are no significant differences in terms of acid taste among wines acidified with different acidifiers (Gomez et al. 2015).

In this process, the side products formed, potassium sulfate and calcium tartrate, represent two major potential drawbacks:

– Potassium sulfate remains in the wine and exceeds legal limits of sulfates applicable in most countries. Furthermore, high sulfate levels are detrimental to quality since they elicit a 'hard' aftertaste and a 'blunt' sensation on the teeth.

– Calcium tartrate tends to precipitate, but its depletion is delayed and requires several months at least. Thus, high levels of residual calcium can affect taste properties (Chapter 2.2 and 8.4), cause post-bottling crystallization, and require cumbersome measures to achieve calcium stability (Chapter 12.2).

Taking into account these disadvantages and the additional difficulty of dissolving gypsum in must or wine, its use has been largely abandoned and almost entirely replaced by solely tartaric acid additions.

The issue of safe pH

An everlasting discussion in many segments of the wine industry is about the pH considered to be 'safe'. Some very conservative schools of thought keep on advocating to lower pH to 3.5 for safety reasons and to add as much tartaric acid as necessary to achieve this goal. As a consequence, many winemakers are terrified by higher pH figures because they mean more risk of microbial activity since SO_2 is less effective. This is absolutely true.

Other winemakers have learnt to handle high-pH wines by using modern techniques that were not available some decades ago. They feel quite comfortable bottling particularly their high-end red wines with pH in the range of 3.7 to 3.9, stating that wines acidified according to pH taste harsh and tough, with tannins in reds being coarse instead of mellow. This is also true. Many of the great red wines of the world be considered undrinkable by contemporary quality standards if their pH was lowered to 3.5. Unpleasant wines that have been distorted by over-acidification just for pH concerns are easy to find.

Producing outstanding wines quite often involves some risks. Thinking only about analytical figures and microbial safety is not the best way to obtain great wines. Other beverages set an example of how to successfully ensure stability in a high pH environment. The basic techniques available for microbiological wine stabilization are physical ones. They comprise control of oxygen uptake, filtration, and sterile bottling. One should make use of them instead of relying too much on impetuous acid additions. Careful filtration is less harmful than disharmony on the palate.

It is common knowledge that microbial stability in wines not submitted to sterile bottling is obtained by the interaction of pH and free SO_2. These two parameters determine the level of molecular SO_2, which is the microbiologically active form of free SO_2. Hence, once one knows the free SO_2 and pH, the molecular SO_2 can be calculated or obtained from tables or graphs such as figure 3. Depending on authors and textbooks, 0.5 or 0.8 mg/L molecular SO_2 are required for what is considered microbial safety. The whole wine industry worships these values more than any bible, and many wineries are keen to obtain them by rash pH adjustments regardless of the sensory outcome.

Unfortunately, serious problems arise in the determination of free SO_2 in red wines. Conventional methods such as iodine titration, the aeration-oxidation method, and flow-injection analysis using para-rosaniline deliver falsely elevated results in these wines. This increase in apparent free SO_2 is caused by the partial dissociation of the anthocyanin-bisulfite complexes during the analysis. Thus, free SO_2 levels as measured by the widely used aeration-oxidation method are overestimated by 24-76 % in red wines due to the interference of the anthocyanin-bound SO_2. This finding has far-reaching consequences: When free SO_2 data are wrong, molecular SO_2 data calculated therefrom must also be wrong. For this reason, headspace measurements were proposed as an alternative method (Coelho et al. 2015).

Furthermore, the validity of microbial protection provided by molecular SO_2 data has been reevaluated. Thus, it has been shown that it must actually exceed 2.0 mg/L to markedly inhibit the growth of Saccharomyces cerevisiae strains such as EC 1118 (Howe et al. 2018). This is three to four times more than the 0.5 or 0.8 mg/L molecular SO_2 required by traditional teaching.

Sure, Saccharomyces cerevisiae control by SO_2 is not an issue with dry reds. However, these results are very indicative of the elevated free SO_2 levels as measured by techniques such as aeration-oxidation that are required to control spoilage yeasts such as Brettanomyces.

These findings have brought down an enological doctrine that is more than half a century old. Ultimately, they suggest that most attempts to lower pH to levels considered 'safe' have in reality failed in doing so. Nevertheless, hardly any of these wines has become microbially flawed. After sterile bottling wines have no possibility to do so at all. Hence, the widespread pH fetish, predicated by a perception of necessary conditions for molecular SO_2 to facilitate microbial security, should be seriously questioned.

Acid additions recapped

– Added tartaric acid does not remain in solution, but precipitates for the most part with potassium as potassium bitartrate.

– By doing so, it decreases pH more than other organic acids added in equivalent amounts. 1 g/L tartaric acid lowers pH by 0.15 to 0.20 units, depending on the buffer capacity of the must or wine.

– Other acids such as malic, lactic, fumaric, and citric acid in amounts equivalent to 1.0 g/L of titratable acidity taste equally sour, do not exhibit any distinguishable taste properties and can be replaced one by another, for example 0.9 g/L of malic acid by 1.2 g/L of lactic acid.

– Relying on low pH figures only for safety reasons is rarely the best way to produce great wines.

4.

Acidification by physical means

The physical means of acidification comprise ion exchange and electrodialysis. Their use is subject to variable national jurisdictions and limited to large-scale operations to be economic. On the other hand and on the grounds of ethical and political aspirations to dispense with additives to must and wine, they may also become more interesting in smaller winery settings.

4.1. Ion exchange

Cation exchange resins are approved for acidification in most jurisdictions. They consist of insoluble, polymeric resin beads containing functional groups capable of binding mineral cations, mainly potassium ions, and to replace them with hydrogen (H^+) ions. They are also frequently found in homes as water-softening devices. In this case the metals targeted are the divalent metals calcium and magnesium, which are exchanged for soluble sodium ions on the resin.

Depending on the kind of the functional group (sulfonic or carboxylic), cation exchange resins show strong or weak exchange capacities. For removal of potassium and other lower charge state cations, strong exchangers are preferable. When they are loaded with H^+ ions by washing them with a strong acid prior to use, they will exchange K^+ and, to a minor extent, other cations such as calcium and magnesium for H^+ ions during treatment of wine. The increase of the H^+ ion concentration lowers pH.

The removal of mineral cations converts salts into free acids, thus increasing TA. This reaction goes on until the total exchange capacity of the resin, expressed as charge equivalents per liter of resin (eq/L), is depleted.

Based on charge and atomic weight of the cation to be removed, the amount of resin required to remove x mg/L of potassium (or any other cation) from a given wine volume can be calculated. In order to avoid the need for precise calculations, the wine industry tends to treat only a portion of the wine or juice with an excess of exchange resin. The treated portion, in which usually more than 75 %

of the mineral cations is removed resulting in a decrease of pH to near 2, is then tested and blended back into the untreated wine.

Theoretically, prepared resins can be used in batch and added directly added to the wine. However, their use in a continuous flow-through process using pre-packed cartridges filled with the resin is more frequent and economic. In both cases, however, they have a detrimental effect on wine quality because they are not absolutely specific and pull out some other wine components such as phenolic and aromatic compounds by non-specific adsorption on the resin matrix. The sensory outcome is difficult to predict.

One of the ways to mitigate undesirable side effects on wine quality is treating only a portion of the wine. Another approach makes use of a kind of nanofiltration operating with a membrane that rather specifically lets mineral cations pass through into the permeate. Only the permeate containing most of the wine's mineral cations is pumped through the cation exchange resin cartridge and then fed back into the retentate. Other wine constituents - aromatics, phenols etc. responsible for the wine character remain in the retentate and do not contact the exchange resin.

The regeneration of these cation exchangers is carried out with concentrated sulfuric acid solution. Much water is required for subsequent rinsing. Effluents must be treated and recycled as a special waste, making this technology quite expensive from the point of view of environmental sustainability.

In contrast to cation exchange resins, anion exchange resins are not allowed in the EU despite their possible use, alone or in combination with cation exchangers, to correct wine acidity. This is due to the potential release of traces of quaternary ammonium salts able to modify wine sensory characteristics.

4.2. Electrodialysis

Electrodialysis is another approach to decreasing ionic compounds. It removes ions from a solution by using ion-selective membranes and the application of a DC voltage.

Acidification as well as deacidification by electromembrane techniques derive from electrodialysis used for bitartrate stabilization (Section 9.2). Classic electrodialysis provides the separation of cations and anions from wine by means of an electric field and a membrane pack, where cationic and anionic membranes are alternately assembled. Cationic membranes carry sulfonic functional groups ($-SO_2O^-$) and are permeable only to cations, whilst anionic membranes are functionalized with quaternary ammonium groups ($-NR_4^+$) allowing the diffusion of anions only. When cations and anions move toward the opposite poles of the electric field applied, they are extracted from wine and concentrated in water (brine) that circulates, inside the membrane pack, in adjacent compartments with respect to those where wine flows.

A modification in the assembly of the membrane pack allows this technique to also be used for acidification and deacidification purposes. Therefore, bipolar membranes are used. They are formed by laminating together cation- and anion-exchange membranes through an intermediate junction layer. Thus, these membranes have both a cationic and an anionic face, so that they do not allow the permeation of either cations or anions.

When used for acidification, bipolar membranes are coupled with cationic membranes. The wine circulates inside the membrane pack, between cationic membranes and the cationic side of the bipolar ones. Water flows in the adjacent compartment. When the electric field is applied, potassium cations move toward the cathode and cross the cationic membranes. Thus, they are extracted from wine and concentrated in the water, which turns to brine. In the wine compartment, potassium is replaced with the protons (H^+), which are formed at the bipolar membrane junction. By analogy, bitartrate anions tend to move toward the anode, but they are forced to remain in the wine, because they are unable to cross the cationic layer of the bipolar membrane. As a result, the wine retains bitartrate and anions of other organic acids, while it is enriched with H^+ ions. Consequently, the pH decreases and total acidity increases. Simultaneously, the water used at the beginning of the process becomes gradually more concentrated in K^+ and OH^- ions, thus becoming brine (Comuzzo and Battistutta 2018).

Bipolar membranes are the most recent outcome of continuing developments in applying membrane technology to wine (Rozoy et al. 2013, Rayess and Mietton-Peuchot 2015), in particular to bitartrate stabilization (Section 11.2). They allow a very precise adjustment of pH regardless of buffer capacity or bitartrate precipitation in a single passage and under completely automatized conditions. Their drawback is their high demand on water treatment and consumption. In a nutshell, these new membrane technologies are just at the onset of their development and can be expected to undergo much improvement in the future.

Cation exchange and electrodialysis are subtractive measures. They remove mineral cations and thus convert neutralized acids into their respective free forms. The ratio of neutralized to free, titratable acid depicted in figure 1 shifts in favor of free acid without changing total acidity (Chapter 1.1). The use of both techniques is limited to medium to large scale wineries. In small and medium sized wineries, additive measures of acidification by adding acids are less cumbersome.

Acidity Management in Musts & Wines

5.

Acidification by biological means

Saccharomyces cerevisiae yeast strains form some organic acids during alcoholic fermentation, which in total increase the titratable acidity by 1 to 2 g/L. However, this gain is commonly cancelled out by the acidity loss that results from bitartrate precipitation during and after fermentation. Non-Saccharomyces strains such as Lachancea thermotolerans are able to induce effective acidification by forming several g/L of lactic acid from sugar, but so far they are hardly disseminated in commerce and winemaking. Acidification by blending with must or wine obtained from unripe grapes resulting from early cluster thinning is considerably detrimental to quality due to the presence of unripe aromas and tannins, unless such a blend wine is sensorially neutralized by fining with significant amounts of charcoal in such a way that only the acids remain.

5.1. Acidification by blending with wines from unripe grapes

The simplest way of increasing wine acidity is the blending of must or wine with products obtained from low-maturity grapes. The possibility of producing a high-acidity must, essentially characterized by a high malic acid content, starts with grapes harvested during early cluster thinning at the beginning of veraison in the vineyard. This practice may be useful when pursuing a low-impact winemaking approach without additives as is the case in organic winemaking. As an example, when unripe grapes are harvested at the proper moment, their juice can contain circa 35 g/L TA. Hence, the addition of only 3 % of that juice increases TA by 1 g/L.

Taking into account the percentage of cluster thinning usually applied in the vineyard, the resulting juice is largely sufficient to correct the acidity of all the wine produced in the same vineyard. In this way, a viticultural by-product is valorized (Celotti et al. 2007).

Unfortunately, musts and wines obtained from low-maturity grapes display green and vegetal aroma characteristics, as well as very aggressive and harsh tannins in the case of red wine. These characteristics of immaturity are easily transmitted to the total quantity of wine after blending. The same consideration applies to blending the grapes instead of must or wine. Therefore, such practices should be carefully managed.

In order to reduce the sensory problems other than high acidity associated with wines obtained from unripe fruit, the wine obtained from cluster thinning can be treated with high amounts of charcoal (5 g/L) and bentonite (1 g/L) to obtain an odorless and colorless wine. When such a wine with 17.8 g/L TA was used to reduce pH and ethanol content of different red wines produced from grapes harvested at complete phenolic maturity, the acidified wines showed similar phenolic composition and sensory characteristics with respect to their controls. Thus, this procedure allows the production of wines with increased acidity, reduced alcohol content, and similar sensory properties without specific equipment. Ultimately, it offers a means of addressing the typical problems of over-ripening (Kontoudakis 2011) in times of global climate crisis.

5.2. Acidification by yeast strains

During primary fermentation, all yeasts produce some organic acids as part of their secondary metabolism. However, a resulting increase in total acidity is usually not observed because this synthesis of acids is counterbalanced by a loss of acidity due to potassium bitartrate precipitation. Therefore, young wines usually display less acidity than the corresponding musts. Only in a few cases, when initial titratable acidity and tartaric acid in the must are very low, will the wine after fermentation display more TA than before (Schneider 2005).

Acidification by commercial Saccharomyces yeast strains

Figure 11 shows the net increase of TA produced by 16 widely used commercial *Sacch. cerevisiae* yeast strains in common white musts, which were previously stabilized against bitartrate precipitation by adding CMC and metatartaric acid (Chapter 11.3). The average TA increase over all strains was 1.54 g/L (as tartaric acid), in accordance with other studies (Atanassov and Triphonova 2001). Increasing temperature slightly, but significantly, resulted in increased yeast acid production, whilst initial juice turbidity, pH, and TA were without influence. Thus, acidity synthesis cannot be considered a response of yeast to low must acidity levels, but an intrinsic strain feature (Aragon et al. 1998). Biochemical pathways and the impact of fermentation practices have been described in detail (Chidi et al. 2018).

Figure 11: Total acidity (as tartaric acid) produced by various commercial Sacch. cerevisiae yeast strains (A-P) during fermentation of standard white musts (20.5 Brix) stabilized against bitartrate precipitation.
Means of three musts.

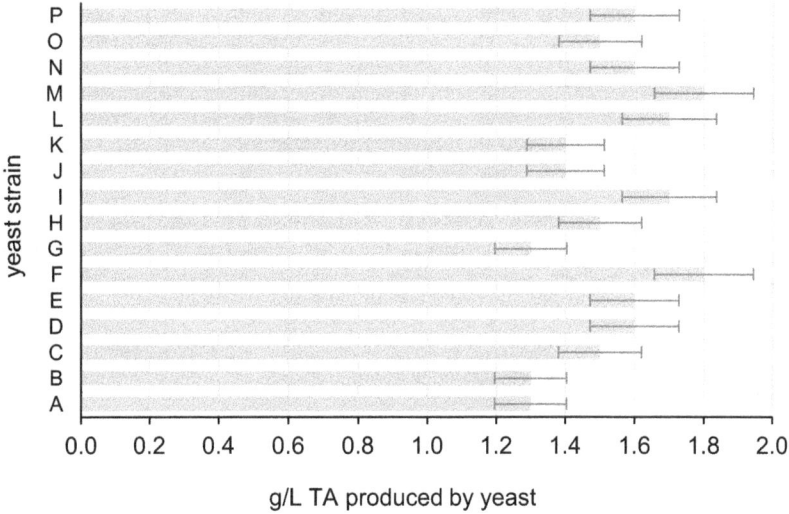

Succinic acid is a normal hy-product of alcoholic fermentation commonly occurring in a concentration range of 0.5 to 1.2 g/L and the most important of the acids produced by yeast. In 93Australian red wines, it ranged from 0.1 to 2.6 g/L with a mean of 1.2 g/L. In 45 white wines of the same origin, it ranged from 0.1 to 1.6 g/L with a mean of 0.6 g/L (Coulter et al. 2004). However, concentrations as high as 3.0 g/L have been recorded in red wines in which strong TA increases have been observed during primary fermentation (de Klerk 2010). It is not yet possible to predict with certainty whether a fermentation will produce a higher than usual amount of succinic acid.

It should be noted that 1.0 g/L succinic acid equals 1.27 g/L TA expressed as tartaric acid. Hence, it can play an important role in wine acidity and be responsible for approximately 50 % of the acids synthesized by yeasts.

Acetic acid is produced by yeasts mainly during their exponential growth phase as a response to hyperosmotic stress conditions at the start of fermentation in amounts of 0.1-0.2 g/L, but in much higher amounts in high-Brix musts such as those used for ice wine production. It strongly correlates with glycerol production. Increased acetic acid production during fermentation has also been associated with a high degree of juice clarification (Delfini and Cervetti 1991), but appears to be more attributable to sugar catabolism by subliminal lactic acid bacteria activity under condition of sluggish fermentations resulting from over-clarified juices.

Citric acid occurs in the metabolism of almost every organism because it is an important intermediate in the tricarboxylic acid cycle. Yeasts produce generally 0.2-0.5 g/L of it (Schneider 2005), in addition to the 0.1-0.2 g/L already present in unfermented juices obtained from sound grapes.

Malic and lactic acid are only produced in negligible amounts by yeasts (Mendes Ferreira and Mendes-Faia 2020) generally not exceeding 0.2 g/L each. More important is the ability of yeasts to degrade malic acid, which relativizes the TA net synthesis of titratable acidity during alcoholic fermentation. Commercial strains of *Sacch. cerevisiae* degrade 0.3-0.7 g/L malic acid (Schneider 2005), but much higher amounts can be degraded by strains of *Schizosaccharomyces pompe* (Chapter 10.2).

In conclusion, it can be stated that yeast strains of *Sacch. cerevisiae* make only a modest contribution to TA, which in most wines is cancelled out by much greater losses of TA due to potassium bitartrate precipitation. Hence, one cannot expect to achieve any effective and reliable acidification with the use of such strains.

Acidification by non-Saccharomyces strains

Some non-Saccharomyces strains can have a positive influence on wine aroma complexity and also produce appreciable amounts of acids during fermentation, but most of them are bad fermenters. To overcome the shortcomings of alcoholic fermentations with such yeasts, the development of mixed fermentations with *Sacch. cerevisiae* acting as an efficient co-fermenter is attracting growing interest.

Lachancea thermotolerans is a non-Saccharomyces yeast strain known for its ability to produce considerable amounts of L-lactic acid by degradation of glucose. When it was used in combined fermentation with a *Sacch. cerevisiae* strain and compared to a single fermentation with *Sacch. cerevisiae* in a white Airén juice, wine quality increased mainly due to the acidification performed by *L. thermotolerans*. It caused a lactic acid increase of 3.18 g/L and a pH decrease of 0.22 units compared to the control fermentation performed by *Sacch. cerevisiae* alone. Sequential inoculation starting with the non-Saccharomyces strain is required for best results with respect to acidity increase (Benito et al. 2016).

In an industry trial with another *L. thermotolerans* strain used in red wine under conditions of sequential inoculation in mixed fermentation with *Sacch. cerevisiae* (EC 1118), *L. thermotolerans* caused an increase of lactic acid by 6.2 g/L, an increase of titratable acidity by 5.4 g/L, and a decrease of pH by 0.16 units. Although the *Sacch. cerevisiae* strain dominated over *L. thermotolerans* during fermentation, the latter showed a high level of competitiveness. It produced also more glycerol, more 2-phenylethanol, and stronger spicy aroma notes under all fermentation conditions tested (Gobbi et al. 2012).

Meanwhile, several strains of *L. thermotolerans* are commercially available. Their suppliers recommend subsequent inoculation with *Sacch. cerevisiae* already after 24 hours to avoid excessive acidification. However, even after inoculation of the *Sacch. cerevisiae* strain, the formation of lactic acid continues for some time. It is also strongly dependent on temperature. In the temperature range from 14 to 22° C, differences of several grams per liter occur. Furthermore, *L. thermotolerans* is poorly SO_2-resistent; the addition of 20 mg/L SO_2 to must largely inhibits lactic acid formation (Burkert et al. 2021). Therefore, the acidification effect is difficult to control in practice. It is recommended to treat only a fraction of the must with *L. thermotolerans* in order to be able to blend it later with the untreated aliquot for correction.

As a result, it can be concluded that co-fermentation with *Lachancea thermotolerans*, and possibly other non-Saccharomyces yeast strains, represents a promising innovation to counter the effect of global warming on grapes and to increase acidity by microbiological means to a practically usable extent. This approach seems particularly interesting to organic or biodynamic growers who are reluctant to use chemical additives. The lactic acid formed is microbiologically and chemically stable.

L. thermotolerans also has potential enological interest from a completely different point of view: It has been shown to be able to decompose acetic acid under aerobic conditions, thus opening a way to develop a controlled biological deacetification process of wines with high levels of volatile acidity (Vilela-Moura et al. 2008)

Acidity increase by genetically modified Saccharomyces strains

Genetic engineering is capable of delivering similar results. Thus, eight genetically engineered *Sacch. cerevisiae* strains over-expressing a bacterial lacticode-hydrogenase enzyme activity were shown to be able to perform mixed lactic acid-alcoholic fermentation. Under enological conditions and depending on must composition, the amount of lactic acid produced ranged from 2.6 to 8.6 g/L and the pH decrease from 0.20 to 0.35 units in comparison with classic alcoholic fermentation. Under industrial conditions, two musts with low or moderate acidity levels were effectively acidified by a 50 % increase in TA and a pH decrease of 0.17 to 0.27 units through the production of 5 g/L lactic acid. Concurrently a reduction of 0.25 % alcohol was observed as a result of the diversion of sugar from ethanol to lactic acid synthesis, while volatile acidity remained unchanged (Dequin et al. 1999).

6.

Principles of chemical deacidification

Chemical deacidification is basically associated with cool-climate growing areas and some high TA-varieties under conditions when biological deacidification by malolactic fermentation is not feasible. It makes use of calcium or potassium carbonates to precipitate a portion of the tartaric acid that might include, under very special operating conditions, also some malic acid. The instantaneous concentration of tartaric acid determines the procedure and the carbonate to be used. Potassium carbonates are preferred, because they facilitate subsequent crystal stabilization, whilst calcium carbonate is only used when large amounts of acidity have to be removed. Whatever the carbonate employed for chemical deacidification, elevated levels of residual potassium or calcium can impact in-mouth sensations, thus demonstrating that this way of deacidification encompasses much more than simply adding a carbonate in an amount proportional to the amount of TA to be removed. Residual potassium can contribute to sensory benefits when potassium carbonates are used in low amounts for sensory fine tuning of almost finished wines, based on previous bench trials. Thus, when chemical deacidification is judiciously performed, it can go beyond rough acidity reductions and substantially improve wines whose TA and pH figures do not suggest deacidification at the first glance.

6.1. Legal aspects

The USA is the only jurisdiction that permits the addition of water to reduce acidity, provided that the wine is obtained from must having a fixed acidity level higher than 5 g/L. Without any doubt, such a procedure changes the character of the wine far beyond a deacidification. However, it is not approved in California, where water is only allowed to be added to high sugar musts with the intent to prevent stuck fermentations. In all other countries, water is merely allowed in negligible amounts as an indispensable processing aid, for example to dissolve fining agents, since there is no quality interest in diluting wine with water. Instead, most legislations permit the use of potassium and calcium salts to lower TA by precipitation of tartaric acid as a means of chemical deacidification.

The USA and Canada permit calcium carbonate, potassium carbonate, and potassium bicarbonate for chemical deacidification. The USA also allows ion exchange for that purpose.

The EU allows chemical deacidification with potassium bicarbonate, calcium carbonate, potassium tartrate and calcium tartrate in all zones except zone C III. Additionally, tartaric acid is permitted for deacidification using the double-salt procedure.

Potassium hydroxide is not approved by any legislation, though there are no enological reasons to prevent its use for deacidification.

6.2. Definitions and procedures

The most frequently used agents for chemical deacidification are carbonates. They comprise potassium carbonate (K_2CO_3), potassium bicarbonate ($KHCO_3$), and calcium carbonate ($CaCO_3$). The effect of the first two is based on reactions of the potassium ions they donate, while calcium carbonate works through its calcium. The carbonate content of these agents indicates that they are salts of carbonic acid. As they dissolve, they ionize into potassium/calcium and carbonate ions, for example

$$K_2CO_3 \rightarrow 2\,K^+ + CO_3^{2-} \text{ or}$$

$$CaCO_3 \rightarrow Ca^{2+} + CO_3^{2-}$$

Next, the carbonate reacts with H^+ ions to yield water and carbon dioxide:

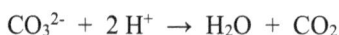

$$CO_3^{2-} + 2\,H^+ \rightarrow H_2O + CO_2$$

The carbon dioxide escapes through the wine surface without any significant effect on the deacidification process.

There are some high purity calcium carbonate salts on the market recommended for the so-called double salt deacidification (Chapter 8.2). They contain a certain amount of already formed crystals of the double salt (calcium tartrate malate) and should be used when this specific approach of deacidification is taken. They are expected to promote its crystallization by acting as crystallization nuclei. Nonetheless, these preparations behave basically as calcium carbonate.

Calcium carbonate can be used in three different ways:

− as standard deacidification removing only tartaric acid;

− as simple double-salt deacidification removing tartaric plus malic acid, the latter in a proportion limited by the tartaric acid concentration;

− as extended double-salt deacidification by adding tartaric acid with the purpose to remove unlimited amounts of malic acid in addition to the tartaric acid.

The potassium ions provided by potassium carbonate or bicarbonate are mono-valent and as such not able to generate a double salt consisting of tartrate and malate.

Any chemical deacidification proceeds in two phases:

1. The neutralization of acids to the corresponding potassium or calcium salts after addition of the deacidification agent. This reaction takes place sponta-neously.

2. The precipitation of the generated salts by means of crystallization. It pro-ceeds slowly and is responsible for long waiting times frequently required after chemical deacidification.

When potassium carbonate or bicarbonate is used, the salt that precipitates is cream of tartar (potassium bitartrate). When calcium carbonate is used, the salt dropping out is calcium tartrate in the case of standard deacidification, and cal-cium tartrate malate in the case of double-salt deacidifications.

What's the best way to deacidify?

There is no general answer to that question. Deacidifications can be carried out by chemical, physical, and biological means. All these procedures are discussed in the following sections. While biological deacidification is a standard proce-dure for almost all red wines and physical deacidification restricted to large op-erations, chemical deacidification is a serious and persistent issue for white musts and wines from cool-climate growing areas, cool seasons, and some high-TA varieties even including some red varieties.

Despite its importance, chemical deacidification appears difficult to master in many wineries. Hence, it is one of the primary topics in the following sections. One of the problems to resolve is the choice between calcium and potassium carbonates. The answer depends on a large array of analytical features and the sensory outcome one is looking for. Even though both carbonates can replace each other in certain situations, they have quite opposed effects on other wines where the choice must be clear and unequivocal.

Some commercially available proprietary blends containing unspecified propor-tions of both calcium and potassium carbonates promise easy deacidification regardless of the individual wine composition and style. Their careless use can entail disastrous consequences and is highly discouraged.

6.3. The importance of tartaric acid content

Tartaric acid is the only acid in musts and wines that is able to generate insoluble salts with both calcium and potassium ions provided by deacidification agents. Therefore, its content determines the maximum extent of chemical deacidifica-tion. Even when a double-salt deacidification is carried out, malic acid can only be removed in conjunction with tartaric acid. Therefore, the removal of malic acid and the maximum deacidification span provided by a double-salt deacidification are limited by the concentration of tartaric acid (Chapters 8.2 and 8.3).

Only a part of the tartaric acid is available for deacidification. This portion is the amount exceeding 1 g/L when calcium carbonate is used regardless the way it is applied, i. e. as standard or double-salt deacidification. This means that at least 1 g/L of residual tartaric acid should remain in the wine. The less residual tartaric acid remains, the higher the concentration of residual calcium and the larger the problems resulting therefrom (Chapter 8.4).

When potassium carbonates are used, a residual tartaric acid of 1.5 g/L should be taken into account. This threshold refers to unfiltered white wines at current storage temperatures of $10 \pm 5°$ C $(50 \pm 40°$ F) and is somehow empirical. However, when specific measures for cold stabilization are taken, deacidification calculations can also be based on 1.0 g/L of residual tartaric acid.

Using some simplification that is admissible for technical considerations, one can state that the titratable acidity of musts obtained from sound fruit is made up only by tartaric and malic acid. The relative amount of these acids sometimes leads to conclusions about grape ripeness. This approach is partially justified by the different behavior of both of these acids during grape ripening.

Tartaric acid is biologically almost inert. Compared to malic acid, it undergoes much less degradation during ripening. After crushing or pressing, its content is poorly related to grape ripeness and varies between 4 and 8 g/L at best. In contrast, malic acid is easily depleted in the fruit by respiration. Therefore, juices obtained from ripe grapes show markedly lower levels of malic acid than those from unripe fruit.

Varying amounts of malic acid are primarily responsible for the considerable fluctuations of titratable acidity while tartaric acid remains relatively constant. When the malic acid level is high, TA also tends to be high. In this case, the percentage share of tartaric acid is obviously lower.

Tartaric acid poorly correlates with TA

Many winemakers in cool-climate areas usually in need of chemical deacidification tend to overestimate the accuracy of the general rule outlined above. This can be seen when they try to assess tartaric acid content as a percentage of titratable acidity, assuming that a given TA would correspond to a certain amount of tartaric acid. Or, as an example, that 9.0 g/L of titratable acidity would correspond to 50 % or 4.5 g/L of tartaric acid. This approach easily turns out to be disastrous when it is used to calculate the amount of tartaric acid available for chemical deacidification.

In reality, tartaric acid content is not closely related to TA. Annual weather conditions and differences between growing areas, microclimates and even vineyard blocks make the actual tartaric acid level hardly predictable. Hence, when juices are to be deacidified for more than 2 g/L TA, the choice of the deacidification method should be based on the tartaric acid concentration previously measured. For wine deacidification, knowledge of tartaric acid concentration is required in any case. Since tartaric acid measurement is a standard procedure of contemporary wine analysis, there is no reason to forego and replace it by adventurous estimates.

Even though there might emerge some weak relationship between TA and tartaric acid in juices, it is without any foundation after primary fermentation. The reason for this lies in the heavy and variable losses of tartaric acid, which drops out as potassium bitartrate during and after fermentation (Chapter 2.3). The general tendency is that these losses are all the higher the more elevated the initial potassium content is. Additional factors further complicate this, making the tartaric acid remaining in the wine absolutely unpredictable under practical winemaking conditions. Since quite often too little attention is given to these details, many wines end up entirely different than expected after deacidification.

6.4. Timing of chemical deacidification – must or wine

Without any doubt, minor chemical deacidifications are possible at any stage during production and storage of wine. Frequently, they can even appear useful in the realm of sensory fine tuning when the wines move into their final phase. A typical example is the addition of some 0.3 or 0.5 g/L of potassium carbonate, followed by addition of metatartaric acid or CMC (Chapter 11.3) for crystal stabilization just before bottling. When such minor acidity corrections are carefully performed, they are not associated with side effects detrimental to overall wine quality. On the contrary, they can make a tremendous contribution to in-mouth quality.

However, the question of timing of chemical deacidification is different during bad years in cool-climate growing areas with juice at 10 or 12 g/L TA, which has to be transformed into white wines with 6.0 or 6.5 g/L TA. Therefore, this section is essential for cool-climate producers and the typical fresh, fruity white wines with distinctive varietal character they typically specialize in. The question is whether extensive deacidifications by several g/L TA are better carried out on the juice or on the wine.

Traditional textbooks advocate deacidification of the juice. This advice is justified by better aroma preservation when treatments deemed necessary are carried out prior to primary fermentation. Carbon dioxide bubbles escaping from wine would be able to strip out highly volatile fermentation aromatics that exist in fine white wines. The bubbles cannot do so when they are produced upon deacidification of juice. In reality, the answer is much less unequivocal.

Timing of deacidification affects potassium content and type of wine

As already pointed out, considerable amounts of tartaric acid drop out as potassium bitartrate during and after alcoholic fermentation. With a certain variation, these precipitations halve the initial contents of tartaric acid and potassium in most wines (Chapter 2.3). It should be noted that these proportions only apply when no deacidification has taken place prior to fermentation.

When juice is deacidified, tartaric acid is removed and no longer available for the precipitation of potassium. Each g/L tartaric acid that is removed by juice deacidification stabilizes an additional 262 mg/L of potassium in solution. Deacidifications removing several g/L of titratable acidity from juices are not unusual for certain varieties grown in cool-climate areas. Under these conditions, tartaric acid is reduced to an extent, which prevents any further potassium bitartrate precipitation and stabilizes potassium at its initial high juice level. Postponing deacidification until after fermentation allows for a natural depletion of potassium.

Based on a comparable final TA, the basic difference between juice and wine deacidification lies in the potassium content of the finished wine. The gustatory differences resulting therefrom are far reaching and directly impact the way the wine is perceived on the palate. They explain why wines deacidified prior to fermentation show up as more full-bodied and less sour than could be explained by their TA. They contain more potassium masking sourness (Chapter 2.3). How this effect is rated in terms of quality depends on the individual wine and consumer preferences.

Deacidification of juice vs. wine – benefits and drawbacks

The natural potassium content of juice depends on soil and annual climate conditions. Damp growing conditions during grape ripening mobilize soil potassium and promote its uptake and storage in the grapes. In typical cool-climate growing areas, years with wet weather during ripening frequently yield less ripe fruit, which also has an elevated juice acidity. When juice deacidification is carried out under these circumstances, wines can be flawed by surplus potassium. In analytical terms, they display relatively high pH figures in relation to TA.

High potassium levels can be detrimental to quality and high pH values affect microbiological safety during fermentations. This is a strong argument against deacidification of juice. On the other hand, deacidification of wine also has drawbacks.

While juice can easily be deacidified by standard procedures using potassium (Chapter 7) or calcium (Chapter 8) carbonate, extensive deacidification of wine frequently requires the more cumbersome double-salt procedure (Chapters 8.2 and 8.3). The reason for this are the heavy losses of tartaric acid during fermentation already discussed. Dropped out tartaric acid is no longer available for simple standard procedures of deacidification. Ultimately, the currently available tartaric acid content and the extent of deacidification are the deciding factors for the deacidification procedure to be applied.

When deacidification of young wine requires the use of calcium carbonate in one way or another, measures have to be taken to reduce residual surplus calcium and stabilize the wine against post-bottling calcium precipitations. Adequate procedures for doing so are tricky and time consuming (Chapter 12.2). On the contrary, deacidification of juice allows for a gain of time and natural calcium stabilization until bottling.

Not without justification, double-salting post fermentation is expected to stress the wine and remove lots of precious fruity aromatics during the vigorous stirring of the partial volume required for that intervention (Chapters 8.2 and 8.3). However, collateral damage of that kind can be limited by taking into consideration some technical details.

There is no doubt about aroma losses when a fruity white wine is stirred. Aroma compounds are volatile, can evaporate and be stripped out when carbon dioxide escapes. The extent depends essentially on wine temperature. At temperatures around 0° C (32° F), they are hardly noticeable and confined to highly volatile fermentation-derived aromatics whose short life span does not contribute to long-term wine quality. However, the sensory outcome will be totally different when double-salting is carried out at elevated temperatures on sensitive white wines already affected by previous treatments. As commonly occurs when wines are subject to treatments, the kind of treatment is less important than the way it is carried out.

Technical details to preserve aroma when wines are deacidified with potassium carbonates are given in chapter 7.4.

Working in a winery always means weighing different options since there are rarely enological benefits without disadvantages. Deacidification of juice can give benefits perceived by smell and preserve high potassium levels, while deacidification of wine results in lower potassium levels but may impact aroma. When the intention is to emphasize weight and volume on the palate, one might tend to deacidify the juice and preserve natural potassium. On the other hand, when natural potassium levels are known to be high and when more vibrant and filigree white wines are to be obtained, one might opt for potassium depletion during primary fermentation and deacidifying the wine. In the latter case, measures for preserving the subtle aroma of fruity white wines should be taken. Thus, the question concerning wine or juice deacidification cannot be answered conclusively. Both options have benefits and drawbacks.

In the special case of red wines, chemical deacidification is not usually carried out before primary fermentation since malolactic fermentation is expected to adjust acidity to a satisfying level. However, under cool-climate conditions and after heavy tartaric acid additions to must, this does not happen in all wines. Therefore, an additional chemical deacidification can become necessary after malolactic fermentation (Chapter 10.1) is completed.

6.5. Final titratable acidity after juice deacidification

When white wines are bottled, TA is typically somewhere in the range of 5.5 and 7.0 g/L. When less ripe fruit suggests deacidification of the juice, the question will arise as to the final TA. There is some anecdotal advice to adjust for 7 to 8 g/L TA, leaving a margin for further losses of acidity expected to occur during or after primary fermentation. Possibly, potassium bitartrate precipitation or some subliminal malolactic fermentation will occur.

– Acting on such sort of advice usually leads to wines showing 7 g/L TA or more, thus requiring a second deacidification on the wine, which will be time consuming and not easy to carry out. The reason is that after a deacidification of juice, tartaric acid in the resulting wine is usually depleted to an extent that requires labor-intensive double-salting (Chapters 8.2 and 8.3). For that reason, any deacidification of juice should adjust TA to the desired level in the corresponding wine. There are two reasons why TA after deacidification of juice remains quite stable:

– When juice is not deacidified, TA losses of some 2 g/L usually observed after primary fermentation are the result of potassium bitartrate precipitation (Chapter 2.3). In contrast, juice deacidification by several g/L of TA lowers tartaric acid to an extent that there is no more tartaric acid available for precipitation as potassium bitartrate. Thus, this possibility of losing acidity does not exist anymore.

There is no doubt that increasing pH by early juice deacidification enhances the risk of additional acidity losses by spontaneous malolactic fermentation (MLF). This might occur in some wineries contaminated by aggressive bacteria populations inducing spontaneous MLF almost systematically. In most wineries, however, MLF is an exception in white wines as long as it is not promoted by inoculation. It is hampered by low storage temperatures in cool-climate wineries relying on chemical deacidifications. It is completely inhibited after adjusting free sulfur dioxide in young white wines. An elevated pH as such is not reason enough to make MLF happen. Furthermore, its absolute value is crucial.

In any case, the sensory outcome of juice deacidification is difficult to predict regardless of any incidental malolactic fermentation occurring. This is also true when the desired TA level in the wine is actually achieved. Therefore, it is frequently the case that further acidity corrections appear necessary. In the worst case, they require an extended double-salt deacidification. At best, they can be achieved by minor acidity adjustments upwards or downwards following the guidelines given in sensory exercises 4 and 6.

7.

Chemical deacidification with potassium carbonates

Potassium carbonate and potassium bicarbonate can replace each other when one takes into account their different degrees of efficiency. However, these are additionally differentiated by the extent to which the potassium they provide to the wine precipitates as insoluble potassium bitartrate, which in turn depends primarily on the concentration of tartaric acid currently available in the particular wine. As a result, the efficiency of each of these two products can vary by up to a factor of 2. In red wines, it is additionally influenced by other factors than the acid composition. This makes targeted deacidification with potassium carbonates more complicated than commonly assumed. In order to clarify the relationships and to facilitate computations, some case studies are given.

7.1. Kinetics and efficiency factor

Potassium carbonate (K_2CO_3) and potassium bicarbonate ($KHCO_3$) are legalized for deacidification in most jurisdictions. Their active part, the potassium ions they provide, shows chemical and sensory properties that are quite different from those of calcium ions, which come into play when calcium carbonate ($CaCO_3$) is used for deacidification. Though all these salts are carbonates, i.e. salts of carbonic acid, they show a different behavior and should not be confused.

Both potassium and calcium ions are expected to precipitate in conjunction with tartaric acid, thus lowering the latter and decreasing titratable acidity. However, this reaction is neither spontaneous nor complete. When elevated potassium or calcium levels remain in the wine, they can affect the taste, lead to post-bottling crystal precipitations, and require measures for crystal stabilization.

Potassium is the most important mineral cation in wine, contributing substantially to body and volume on the palate (Chapter 2.3). Thus, an increase of wines' natural potassium content is not necessarily detrimental to quality when potassium provided by potassium carbonates does not completely precipitate. Furthermore, the potassium bitartrate generated can easily be stabilized to avoid post-bottling precipitation (Chapters 11.2 and 11.3). In contrast, elevated calcium residues always affect taste in a negative way (Chapter 8.4) and are much more difficult to stabilize against post-bottling crystallization (Chapter 12.2).

Therefore, potassium carbonates are preferred to calcium carbonate, taking into account the currently available amount of tartaric acid.

In contrast to calcium that is able to also remove malic acid under the specific conditions of the double-salt procedure (Chapter 8.2), potassium supplied by potassium carbonates is a monovalent cation, which can only precipitate tartaric acid as potassium bitartrate. This precipitation is much faster than precipitation of calcium salts and requires some two to three weeks when it occurs spontaneously in young white wines stored at typical cellar temperatures. After this period of time, the relative degree of bitartrate stability or instability existing before deacidification is restored. Absolute crystal stability can easily and quickly be achieved by the usual measures for bitartrate stabilization (Chapters 11.2 and 11.3). Thus, potassium carbonates are also suitable for deacidification shortly before bottling.

Efficiency factor depends on tartaric acid content

When practitioners deacidify with potassium carbonates, they expect the potassium ions to completely drop out with tartaric acid as potassium bitartrate. Thus, according to stoichiometry of basic chemistry, they also expect to need 0.67 g/L of $KHCO_3$ or 0.46 g/L of K_2CO_3 to remove 1.0 g/L TA. This is the popular formula one can find in most textbooks.

Under practical conditions, however, additions of the aforementioned amounts of $KHCO_3$ or K_2CO_3 need considerable time to give the final TA as calculated by this simplified stoichiometry. The reaction is not spontaneous. It is also quite frequently the case that the final TA is never attained even after longer waiting periods. This means that higher amounts of $KHCO_3$ or K_2CO_3 are required than expected.

Definitely, the assumption that deacidification by 1.0 g/L TA requires 0.67 g/L of $KHCO_3$ or 0.46 g/L of K_2CO_3 only applies to a very limited extent. Reactions are much more complex than a simple precipitation of tartaric acid. The explanation lies in what is understood by "precipitable" or "technically usable" tartaric acid. This complicates calculations. The following explanations will illustrate why the efficiency factors of $KHCO_3$ and K_2CO_3 depend considerably on the individual wine, in particular on its tartaric acid content.

Both potassium carbonate and bicarbonate work similarly, but with a different stoichiometry since they differ in terms of potassium content. Thus, they are interchangeable according to the conversion factor

$$1.0 \text{ g/L } KHCO_3 \triangleq 0.69 \text{ g/L } K_2CO_3$$

In the following explanation, details are presented only by the example of $KHCO_3$, which can easily be converted into K_2CO_3.

When $KHCO_3$ is added to wine, the acidic environment makes it dissociate into potassium cations and carbon dioxide according to the formula

$$KHCO_3 \rightarrow K^+ + HCO_3^-$$

$$HCO_3^- + H^+ \rightarrow H_2CO_3 \rightarrow H_2O + CO_2$$

In a first step, the potassium cations neutralize an equivalent amount of titratable acidity. Since TA is expressed as tartaric acid, the following general reaction and mass equation applies:

$2 \cdot KHCO_3$ + tartaric acid equivalents $\rightarrow$ potassium salts

$2 \cdot 100.1$ + 150.1

1.334 g/L $\triangleq$ 1.0 g/L

Hereby, 100.1 and 150.1 are the molecular weights of $KHCO_3$ and tartaric acid, respectively.

This means that the addition of 1.334 g/L of $KHCO_3$ leads to the neutralization of 1.0 g/L TA. The reduction of TA generated thereby is immediate. There will be no additional losses of TA as long as the potassium ions added remain stable in solution. This is the case when there is no tartaric acid available to precipitate them as potassium bitartrate or when that precipitation is inhibited by wine constituents.

In the presence of tartaric acid able to precipitate potassium ions, the whole picture will change. It has been shown in chapter 2.3 that potassium bitartrate is an acid salt, in which only one of the two acid groups of tartaric acid is neutralized. It has also been shown that when 1 g/L of tartaric acid precipitates in the form of potassium bitartrate, this inevitably causes a decrease of TA by 0.5 g/L.

Now, let's imagine that the wine contains enough tartaric acid and a chemical makeup that allow the potassium ions added to entirely precipitate as potassium bitartrate. In such a case, the TA neutralization as the first step is followed by a reduction of TA to exactly the same extent. In other words, the addition of 1.334 g/L of $KHCO_3$ now leads to the reduction of TA by 2.0 g/L in total. As a result, we arrive at the popular formula stipulating 0.67 g/L $KHCO_3$ to reduce TA by 1.0 g/L. This formula already takes into account the loss of TA, which would occur when the added potassium drops out completely. Under these conditions the whole reaction and mass equation may be written as:

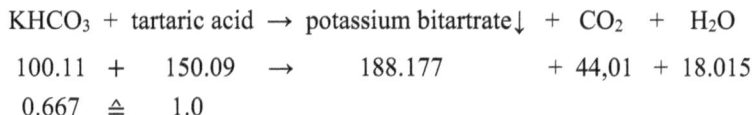

$KHCO_3$ + tartaric acid $\rightarrow$ potassium bitartrate$\downarrow$ + CO_2 + H_2O

100.11 + 150.09 $\rightarrow$ 188.177 + $44,01$ + 18.015

0.667 $\triangleq$ 1.0

This means that 0.67 g/L $KHCO_3$ is required to precipitate 1.0 g/L tartaric acid as potassium bitartrate (KHT).

In other words, there is a two-step reaction:

- Addition of 0.67 g/L $KHCO_3$ $\rightarrow$ immediate reduction of TA by 0.5 g/L

- Precipitation of K^+ as KHT $\rightarrow$ further reduction of TA by another 0.5 g/L

Cumulative effect: 0.67 g/L $KHCO_3$ $\rightarrow$ reduction of TA by 1.0 g/L.

It cannot be overemphasized that the cumulative effect – 0.67 g/L $KHCO_3$ per 1.0 g/L TA – only occurs when the added potassium entirely drops out. Only then is the final TA striven for attained in analytical and sensory terms. This process requires enough tartaric acid and some time for crystallization.

This kind of two-step sequence does not occur when calcium carbonate is used. In this case, calculated final TA is attained right after addition of the $CaCO_3$ regardless of any precipitation of calcium tartrate. As a neutral salt, its precipitation does not affect titratable acidity (Chapter 8.1).

Recap: Efficiency factors of KHCO$_3$

In conclusion, the deacidification with $KHCO_3$ can encounter three different scenarios:

1. Deacidification by 1.0 g/L TA requires 0.67 g/L of $KHCO_3$ when the wine contains enough tartaric acid that can actually be precipitated as potassium bitartrate.

2. When potassium added as $KHCO_3$ is prevented from dropping out due to inhibition of crystallization or lack of precipitable tartaric acid, the reduction of TA by 1.0 g/L requires the addition of 1.34 g/L $KCHO_3$.

3. Only a portion of the added potassium drops out, while the rest remains in solution. In such a wine, the reduction of TA by 1.0 g/L requires the addition of $KHCO_3$ somewhere in the range of 0.67 to 1.34 g/L. In other words, 0.67 g/L of $KHCO_3$ causes a reduction of TA by 0.5 to 1.0 g/L according to the particular wine.

The notions of precipitable and residual tartaric acid

An essential question is about the level of tartaric acid required to entirely precipitate the potassium added as $KHCO_3$. Experience tells us that without targeted cold stabilization, potassium only precipitates with tartaric acid exceeding levels of 1.5 ± 0.3 g/L in standard dry white wines. This threshold can be called residual tartaric acid. The amount exceeding this threshold is referred to as precipitable tartaric acid. In practical terms, this means that when a wine has only 1.5 g/L of tartaric acid, we can assume that almost none of the potassium we add will drop out.

The aforementioned threshold is an empirical one referring to unfiltered white table wines stored at 8 to 15° C (46 to 59° F). It is lowered to approximately 1 g/L when cold stabilization is applied. It cannot be specified more precisely because every wine has its own typical, matrix dependent capability of retaining potassium bitartrate in stable solution at a given temperature (Chapter 11.1).

In red wines, residual tartaric acid is generally higher, variable, and difficult to predict. Many red wines with 2 or 3 g/L tartaric acid do not precipitate any bitartrate after deacidification with potassium carbonates as long as they are not submitted to chilling in conjunction with seeding via the contact process (11.2). In contrast, they do so when they are supplied with tartaric acid in the young stage or as must. It might be hypothesized that red wines develop high-molecular pigments able to act as crystallization inhibitors when they age, whilst these compounds do not occur to that extent in red musts or white wines.

It is important to note in this context that when tartaric acid is measured, this measurement determines the acid anion regardless of whether it exists as free tartaric acid, partially neutralized (potassium bitartrate), or completely neutralized (dipotassium tartrate) salts. Figures 4 shows that in the pH range of wine, the major portion of tartaric acid actually exists in the form of its salts, which are partially soluble. For that reason, wines always display some tartaric acid even after strong deacidifications.

Case studies

The scenarios 1 to 3 mentioned before are clarified by the following real-world case studies:

Case study 1, white wine:

TA = 7.0 g/L

tartaric acid = 2.5 g/L

final TA desired = 6.0 g/L

→ deacidification range = 1.0 g/L TA

residual tartaric acid = 1.5 g/L

→ 1.0 g/L tartaric acid available for precipitation with potassium

Calculation:

   2.5 g/l tartaric acid − 1.5 g/L residual tartaric acid
= 1.0 g/L precipitable tartaric acid x 0.67 = 0.67 g/L $KHCO_3$

Case study 2, white wine:

TA = 7.0 g/L

tartaric acid = 1.5 g/L

final TA desired = 6.0 g/L

→ deacidification range = 1.0 g/L

residual tartaric acid = 1.5 g/L

→ almost no tartaric acid precipitable with potassium.

Calculation:

Deacidification without precipitation of tartaric acid using 1.34 g/L $KHCO_3$. The added potassium will remain in solution and act on the palate regardless of the deacidification achieved.

Case study 3, white wine:

TA = 8.0 g/L

tartaric acid = 2.5 g/L

final TA desired = 6.0 g/L

→ deacidification range = 2.0 g/L

residual tartaric acid = 1.5 g/L

→ tartaric acid partially available for precipitation with potassium

Calculation in 2 steps:

1. Precipitation of 1.0 g/L of tartaric acid with 0.67 g/L $KHCO_3$

2. Neutralization of another 1.0 g/L of TA with 1.34 g/L $KHCO_3$ without precipitation.

This deacidification requires a total of 0.67 + 1.34 = 2.01 g/L $KHCO_3$. The added potassium remains partially in solution and acts on the palate regardless of the deacidification achieved.

Let's summarize once more to clarify: To reduce TA by 1.0 g/L, the amount of $KHCO_3$ required is 0.67 to 1.34 g/L. The exact amount depends on whether and to which extent the added potassium will precipitate tartaric acid. This precipitation does not occur or is incomplete

– when the tartaric acid level currently available is too low to allow precipitation,

– when wines already contain metatartaric acid or CMC (Chapter 11.3) inhibiting potassium bitartrate precipitation,

– in red wines, in which tannins and anthocyanins hamper potassium bitartrate precipitation (Chapter 7.2).

In case study 2 and 3, the specific taste of potassium (Chapter 2.3) will emerge to a greater or lesser extent. Concurrently, there will be a significant increase of pH, which might affect microbiological stability as long as the wine is not filtered.

On the other hand and in contrast to some popular teachings, $KHCO_3$ is not only suitable for minor deacidifications. It can also be used for deacidifying by several g/L of TA bearing in mind the level of tartaric acid that is currently available for precipitation. In contrast to calcium carbonate ($CaCO_3$), its use represents a more gentle treatment of wine since potassium is a natural wine constituent having a positive effect on the palate over a large concentration range. Furthermore and as mentioned previously, subsequent crystal stabilization is easier to perform than after deacidification with $CaCO_3$.

Based on these benefits, deacidification with $KHCO_3$ is preferable to that with $CaCO_3$ as long as the analytical and sensory features of the wine allow for doing so. However, there are typical situations when $KHCO_3$ must be replaced by $CaCO_3$ or other means of deacidification. They occur when

– the initial TA is very high or the deacidification range striven for is high,

– the tartaric acid level is low,

– the pH is already relatively high in relation to TA, thus indicating a naturally elevated potassium content (Section 2.3) that should not be increased any more.

In strictly legal terms, $KHCO_3$ and K_2CO_3 are not considered as approved additives but as treatment agents. Therefore, they should be used in a way that does not enhance wine potassium concentration within the scope of what is possible.

7.2. Peculiarities of red wine deacidification with potassium carbonates

For red wines, malolactic fermentation (MLF) is considered indispensable and commonly used on almost all red wines from virtually all climates and varieties. The reason is that is does not only decrease acidity, but also improves complexity and mouthfeel of the wines (Chapter 10.1). Additionally, the decay of malic acid it causes provides better microbial stability post-bottling. This is an important issue in red wines since in contrast to white wines, they tend to show higher pH figures and are usually stored and bottled with lower levels of free SO_2. As a result, they also have lower levels of molecular SO_2 protecting the wine against detrimental microbial activities (Figure 3). If they still contain malic acid at the moment of bottling and if bottling is not carried out under strict sterile conditions, the risk of post-bottling MLF and cloudiness will considerably increase.

Usually, MLF is deemed sufficient for achieving acidity balance in red wines, but this does not necessarily apply when they are grown under cool-climate conditions or obtained from unusual varieties or unripe fruit. In those wines, high residual TA levels and low pH figures increase the astringency of tannins (Guinard et al. 1986, Fontoin et al. 2008) and lessen the perception of volume and weight. This is the reason why an additional chemical deacidification can become necessary after MLF is completed – a procedure hardly understandable for hot-climate wine growers.

Figure 12 shows the sensory changes on the palate caused by a post-MLF deacidification of a Merlot red wine from 5.7 to 4.7 g/L TA using $KHCO_3$. They comprise much more than only sourness. Since sourness and volume are sensory antipodes, one affecting another, deacidification strongly enhances what is perceived as weight and volume on the palate.

Figure 12: Sensory impact of deacidification of a Merlot red wine from 5.7 g/L to 4.7 g/L TA using 1.2 g/L KHCO$_3$ and cold stabilization.

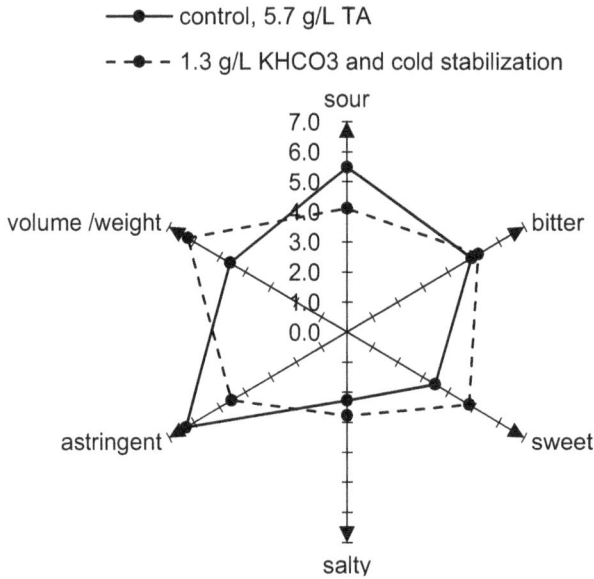

The amount of 1.3 g/L KHCO$_3$ to reduce TA by 1.0 g/L in this red wine requires an explanation: In contrast to white wines, red wines contain high amounts of tannins and anthocyanins that are more or less polymerized and associated with polysaccharides. These molecules can reach a colloidal state, in which they act as protective colloids. Thus, they partially or entirely inhibit the precipitation of potassium bitartrate produced by deacidification with KHCO$_3$. The extent of that effect depends on the particular wine and its current status. It is greater in aged red wines with higher level of colloidal material than in young red wines. It is almost absent in red musts, in which less tannin is present or in which the tannin is not yet in a colloidal state able to inhibit bitartrate precipitation. Hence, when bitartrate is produced by adding tartaric acid to a red must, it rapidly precipitates, but this does not happen to the same extent when it is produced by adding potassium carbonate to a red wine.

Based on the preceding discussion, when red wines are deacidified, the potassium added as KHCO$_3$ quite often remains entirely in solution, even if they contain enough tartaric acid able to precipitate it. Even subsequent rigorous cold stabilization (Chapter 11.2) does not necessarily change this much. In these cases, deacidification by 1.0 g/L TA requires 1.34 g/L of KHCO$_3$. This leads to case study 4 in continuation of the previous case studies on white wines:

Case study 4, red wine after malolactic fermentation:

TA = 6.0 g/L

tartaric acid = 3.3 g/L

final TA desired = 4.7 g/L

→ deacidification range = 1.3 g/L

residual tartaric acid = approx. 3.0 g/L(red wine!)

→ Almost no tartaric acid precipitable with potassium because tannins and anthocyanins inhibit potassium bitartrate crystallization to a variable degree.

Deacidification (almost) exclusively via neutralization using 1.34 g/L $KHCO_3$ per 1.0 g/L TA.

➢ 1.34 x 1.3 = 1.74 g/L $KHCO_3$ to lower TA by 1.3 g/L.

➢ Potassium remains largely in solution and acts on the palate (soapy?)

➢ For safety, subtract 10-20 % from KHCO3 calculated in this way to account for some minor bitartrate precipitation and additional TA losses associated therewith.

➢ Deacidification with $CaCO_3$ could be the better solution when initial potassium levels are high.

However, in the course of further storage or upon cold stabilization, a minor amount of the added potassium will probably precipitate. The additional acidity losses associated therewith are most often modest or even negligible on the palate. On the other hand, they are quite unpredictable since the well-known laws governing potassium bitartrate stability in white wines (Chapter 11.1) do not apply.

Sensory trials on red wines are particularly useful

Tannin quality can significantly impact the perception of sourness regardless of pH and TA figures (Schneider and Tracey 2021). Hence, for any kind of red wine, preliminary tests of deacidification according to the protocol outlined in sensory exercise 6 (Section 7.3) have proven to be extremely useful. When they are carried out, the amount of $KHCO_3$ considered sensorially optimal is paramount and much more important than the final TA and pH figures achieved.

Trials of that kind can disclose additional potential for flavor improvement even in red wines previously judged as perfect. They help better understand the complex relationship between sensory properties and analytical makeup. In numerous red wines, meticulous deacidification provides much more benefits for quality improvement than the popular range of traditional treatments. One of the reasons for this is that, as mentioned previously, mitigating sourness reduces astringency and enhances volume. Increased potassium levels additionally contribute to that effect (Chapter 2.3).

The taste profile achieved by the addition of a given amount of $KHCO_3$ can remain stable over quite a long period of time. However, additional delayed TA losses caused by some slight and unpredictable bitartrate precipitation cannot be excluded in the long term. Therefore, it is beneficial to freeze the trial samples overnight before sensory evaluation takes place. As an alternative and somewhat empirical approach, a delayed TA decrease should be anticipated by reducing the amount of $KHCO_3$ deemed optimum by 20 to 30 %. These precautions do not apply when metatartaric acid is added shortly after $KHCO_3$ additions to prevent further bitartrate precipitations as it is current practice in simple red wines designed for rapid consumption.

Since precipitation of potassium added to red wines is largely inhibited, this way of deacidification can cause a considerable increase of potassium concentration. As a consequence, a sharp increase of pH to 3.7 or more can be observed. This causes microbiological instability when storage temperature exceeds some 10° C (50° F). Therefore, red wines after chemical deacidification should not be aged any longer in barrels but stabilized by filtration, cool storage, or SO_2 addition in a timely manner.

Not all red wines can be deacidified with potassium carbonates

High potassium concentrations can be distracting on the palate regardless of the final TA achieved (Chapter 2.3). This happens frequently when TA of cool-climate red wines is lowered by more than 1.5 g/L or by using more than 2.0 g/L of $KHCO_3$. High initial potassium concentrations additionally amplify that kind of risk. In such circumstances, deacidification with calcium carbonate is more beneficial.

The mechanism of standard deacidification with $CaCO_3$ relies on the plain precipitation of tartaric acid as neutral calcium tartrate (Chapter 8.1). It is important to note that double-salt deacidification with $CaCO_3$ (Chapter 8.2) cannot be used on red wines after MLF since malic acid has been depleted. Fortunately, red wines tend to contain more tartaric acid that can be removed with $CaCO_3$ than whites; their tannins hamper its depletion by bitartrate precipitation.

Additionally, precipitation of calcium tartrate is much less inhibited by red wine tannins than potassium bitartrate precipitation. The higher pH levels in red wines even favor its precipitation. Nevertheless, time requirements and accompanying measures for calcium stabilization should be taken into account in the same way as they apply to white wines after deacidification with $CaCO_3$ (Chapters 12).

When certain red varieties grown under cool-climate conditions are empirically known to require severe chemical deacidification after MLF completed, it can be wise to apply some 0.7 g/L $CaCO_3$, corresponding to a TA reduction of 1.0 g/L (Chapter 8.1), at an early stage to gain time for natural calcium stabilization. Sensory fine tuning using low amounts of $KHCO_3$ can be done at a later time when the finished wine can sensorially be assessed.

The sour taste of red wines correlates less with titratable acidity than in white wines. In other words, TA and pH of red wines are a treacherous index for quantifying perceived sourness. The major reason of that is found in tannins, which is not present in white wines.

Acidity and tannin interact on the palate. Tannin originating from ripe grapes (high phenolic ripeness) displays a sweet flavor component that can feign in some cases 2 to 3 g/L of residual sugar more than the wine actually has. Inversely, tannin extracted from unripe fruit (poor phenolic ripeness) bears a sour sub-taste that is quite able to feign 0.5 g/L TA more than one measures (Schneider and Tracey 2021). As a result, even red wines with less than 5.0 g/L TA can be deemed too sour and require some deacidification. Figure 13 shows schematically how tannin quality and other wine constituents affect sweetness and sourness.

Figure 13: Sensory interaction of tannin and acidity: Impact of tannin and other red wines constituents on perceived sourness.

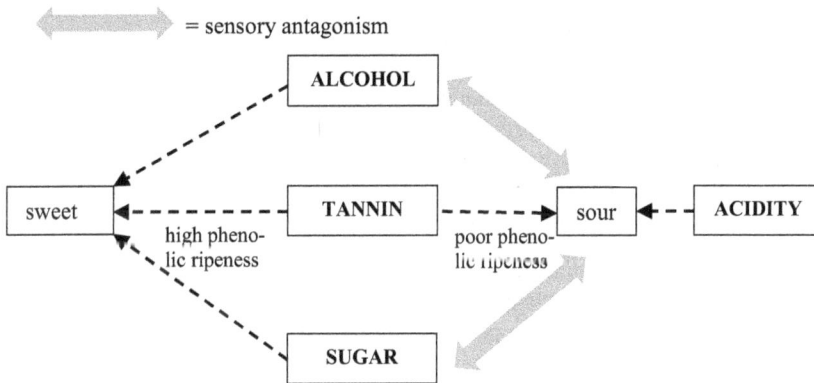

The chemical foundations of this variable gustative effects exerted by tannins are far from being completely investigated. Tannin interactions with anthocyanins and mannoproteins play a role.

7.3. Implementation of sensory trials for wine deacidification

Under real-world conditions, the question of whether a minor deacidification might be useful frequently arises shortly before bottling, when the wine to be bottled is compared sensorially with other ones. In such a situation, the use of $CaCO_3$ is excluded anyway since it would involve time-consuming calcium crystal stabilization. The use of potassium carbonates opens a way to minor acidity adjustments downwards in the short term. Trials allow for sensory optimization and are conducted according to sensory exercise 6.

Sensory exercise 6: Preparation of test solution for deacidification of wine and implementation of trials.						
Prepara-tion of test solution	Dissolve 100 g of $KHCO_3$ and adjust to 1000 mL with distilled water.					
Use of the test solution:	**0.1 mL / 100 mL wine equals in the tank:** + 0.1 g/L $KHCO_3$					
	+ 0.25 mL	+ 0.50 mL	+ 0.75 mL	+ 1,00 mL	+ 1.25 mL	+ 1.50 mL
Equals in the tank:	+ 0.25 g/L	+ 0.50 g/L	+ 0.75 g/L	+ 1.00 g/L	+ 1.25 g/L	+ 1,50 g/L

When these trials are run, one should observe the following points:

− When working on red wines or on wines already containing metatartaric acid, CMC or containing not more than approximately 1.5 g/L of tartaric acid, the trials can be evaluated immediately after the addition of $KHCO_3$. The sensory profile on the palate will remain stable. There will be no further significant decrease of acidity and perceived sourness on the basis of potassium bitartrate precipitation. In this case, addition of 0.67 g/L of $KHCO_3$ reduces TA by 0.5 g/L. See sections7.2 for details, in particular case study 4.

− If a white wine contains more than 1.5 g/L tartaric acid, a belated additional decrease of TA must be expected owing to the precipitation of potassium bitartrate. To accelerate this process, samples should be frozen overnight or stored in the refrigerator during several days before sensory evaluation is carried out. Exceptions apply when the wine has already been stabilized with metatartaric acid or CMC against potassium bitartrate crystallization, or when it will be stabilized shortly after addition of $KHCO_3$.

Learning from the acidification trials outlined in sensory exercise 6, every wine-maker would be well advised to keep a $KHCO_3$ or K_2CO_3 test solution at hand. Its frequent use will help understand why minor deacidifications are a powerful tool of sensory optimization, much more powerful than a large range of commercial additives usually recommended for that purpose. The allusion to the increase in pH and associated microbial risks is invalid, as powerful techniques for filtration prior to bottling are available. Furthermore, despite the caution about elevated pH values, even at 'typical' pH levels, and contrary to near universal belief, pH has never provided microbiological stability in unfiltered red wines. The reasons are explained in chapter 3.5.

7.4. Practical application of potassium carbonates

Potassium carbonate or bicarbonate is primarily used for minor acidity corrections in almost finished wines. Frequently, these deacidifications lead to a removal of only fractions of a g/L TA. The desirability of such minor corrections cannot have been anticipated at earlier stages of winemaking. Usually, they result from the sensory picture of the wine at a moment when it is further developed. For the purpose of sensory fine tuning, they are often performed shortly before bottling.

Potassium carbonates react regardless of the way they are applied. In contrast to calcium carbonate (Section 8.1), they do not undergo any crystal aging with loss of reactivity. However, some commercial brands tend to clump. The complete dissolution of any clumps must be controlled. When this is done properly, differences in the acidity reduction achieved are exclusively due to the variable extent of potassium bitartrate precipitation (Section 7.1).

Despite the easy use of potassium carbonates, they should not be directly added to the whole wine volume. The carbon dioxide produced can cause overflow when no headspace is available in the container. Its release strips out appreciable amounts of volatile aromatics from sensitive white wines with distinctive varietal aroma, particularly when they are deacidified at higher storage temperatures. Red wine aromatics are less volatile, but they become distracted by residual carbon dioxide. For removal of excessive carbon dioxide, pumping-over once or several times (Schneider and Tracey 2021) or adequate membrane contactors (Schonenberger et al. 2014) can be used.

Dissolution in a partial volume is advisable

Another way to avoid the aforementioned side-effects is the dissolution and neutralization of the potassium carbonates in only a fraction of the total wine volume. After blending back the overdeacidified fraction, no more carbon dioxide is produced in the original lot. The volume of the fraction used to dissolve the potassium carbonates must be large enough to provide sufficient acidity for its complete neutralization and the release of all carbon dioxide. Example:

5,000 L of wine with 8.0 g/L TA should be deacidified using 1.5 g/L of $KHCO_3$, corresponding to a total amount of 7.5 kg $KHCO_3$. Complete neutralization of 1.5 g $KHCO_3$ requires 1.5 : 1.34 = 1.12 g TA (as tartaric acid) (Chapter 7.1). This results in 5,600 g TA for 7,500 g $KHCO_3$. In order to make this amount of TA available, the fraction must contain at least 5,600: 8.0 = 700 L.

A modification of this procedure is to put the solid potassium bicarbonate, free of clumps, into a tank and to add the wine slowly making sure the tank is filled from the bottom. This can be done when racking, blending, or filtration require a transfer from one tank to another. After transfer of some 20 % of the total volume, the wine flow is interrupted for some minutes. Formation and release of surplus carbon dioxide takes place in the first subsets of wine flowing into the tank. The transfer replaces vigorous stirring, which would otherwise become necessary to mix the wine after deacidification.

7.5. Promoting potassium bitartrate precipitation during deacidification

Chemical deacidification means basically precipitating tartaric acid in the form of its poorly soluble potassium or calcium salts. For that reason, deacidification invariably affects tartrate stability and requires close attention to measures achieving crystal stability.

The traditional way of deacidification with potassium carbonates is that the de-acidification agent is poured into the whole wine (juice) volume at current cellar temperature, followed by mixing the wine. In order to accelerate the precipitation of the potassium bitartrate produced and shorten the waiting times usually required for crystal stabilization, several practical hints to improve the traditional procedure of deacidification have been proposed (Friedrich and Müller 1999). They are based on the following fundamental laws of crystallization:

– The crystallization rate increases when the temperature decreases.

– The crystallization rate increases when the extent of supersaturation of the respective salt increases. Supersaturation means that there is more salt held in solution than it is possible to dissolve since its crystallization is temporarily inhibited (Chapter 11.1).

– The crystallization rate also increases when the number of crystals in the solvent increases, be it by formation via chemical reaction or by their addition.

– It increases even further when the crystals are brought in close contact with the supersaturated solution.

– It is also affected by pH. Maximum precipitation rate of potassium bitartrate occurs around pH 4.1, corresponding to the pH where the bitartrate fraction of tartaric acid is highest (Figure 5).

In order to take into better account these fundamentals, the traditional deacidification procedure – adding the deacidifying agent to the wine – should be subject to the following modifications:

– First put the deacidifying agent into an empty container. Subsequently, slowly add the wine at reduced pump speed, filling the container from the bottom. Additionally, mix the wine preferably using a propeller mixer screwed onto a tank fitting. This way of working ensures that potassium bitartrate is formed from the very first beginning at a highly supersaturated concentration, which accelerates its precipitation (Chapter 11.2).

– A similar effect can be obtained when a partial fraction of the total volume is overdeacidified and blended back. The smaller the overdeacidified fraction, the more favorable the conditions for rapid crystallization, i.e. maximum concentration of potassium ions.

– Previous cooling of the wine to a temperature of less than 10° C (50° F). The low temperature leads to an additional increase of supersaturation that further promotes crystallization.

– Keep on mixing continuously or at intervals over several hours. Agitating the wine ensures close contact of the crystals already formed with the supersaturated solution. Thereby, the crystals act as contact crystals promoting further crystallization according to what happens during the contact seeding process (Chapter 11.2). If the crystal mass is not held in suspension by mixing, the crystallization rate will dramatically decrease. Therefore, speed and intervals of mixing must be adjusted in a way that prevents crystals from sedimentation. The frequency of mixing intervals determines the crystallization rate and time requirements for crystal stabilization following deacidification.

Theoretically, the aforementioned measures should provide perfect potassium bitartrate stability right after deacidification. In practice, they do not do so with absolute certainty. They do not dispense with methods of controlling bitartrate stability and targeted stabilization prior to bottling. However, they are valuable tools for making subsequent crystal stabilization procedures much easier.

On the other hand, the long stirring periods required cause a strong oxygen uptake, which is particularly feared in sensitive, fruity white varietal wines for the oxidative aging reactions it induces. Furthermore, the release of carbon dioxide strips out volatile aromatic compounds that are highly appreciated in those kinds of wines. Therefore, many winemakers reject these procedures of promoting bitartrate stabilization and rely on the use of crystallization inhibitors (Chapter 11.3) when they feel compelled to deacidify wine instead of must.

7.6. Deacidification with other potassium compounds

As outlined previously, potassium carbonate (K_2CO_3) and potassium bicarbonate ($KHCO_3$) can be replaced one by another. However, basic stoichiometrics explain why they exhibit a different deacidification power in the same wine. The reason is that K_2CO_3 contains more potassium than $KHCO_3$. Thus, 1.0 g K_2CO_3 equals 1.457 g $KHCO_3$, while 1.0 g $KHCO_3$ equals 0.686 g K_2CO_3 (Section 7.1).

Beyond the two potassium carbonates, other potassium compounds can also provide the potassium ions required for deacidification. However, they are rarely used or even illegal.

Potassium hydroxide (KOH) for deacidification purposes has not been approved anywhere, though there is no reason not to use it from a purely enological point of view apart its potentially hazardous handling. In contrast to the potassium carbonates, as a side effect it releases water instead of carbon dioxide when added to wine. This can be beneficial when an enhancement of carbon dioxide in wine is not desirable as is the case with red wines in particular. The amount required to lower TA by 1.0 g/L is 0.374 g/L KOH in the case that the potassium supplied precipitates completely with tartaric acid, and 0.748 g/L KOH when it remains as soluble in the wine.

Dipotassium tartrate ($K_2C_4H_4O_6$), abbreviated as K_2T in this chapter, is the neutral potassium salt of tartaric acid and not to be confounded with potassium bitartrate (KHT). It is approved for chemical deacidification in most wine growing countries of the world, but rarely used because of its high price.

In the acidic environment of wine, it dissociates and reacts with the wine's tartaric acid (H_2T) according to the mass formula

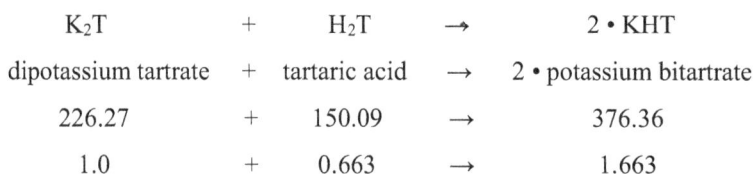

K_2T	$+$	H_2T	$\rightarrow$	$2 \cdot KHT$
dipotassium tartrate	$+$	tartaric acid	$\rightarrow$	$2 \cdot$ potassium bitartrate
226.27	$+$	150.09	$\rightarrow$	376.36
1.0	$+$	0.663	$\rightarrow$	1.663

As a neutral salt, its addition to wine initially does not affect titratable acidity at all. However, when the KHT produced drops out entirely, addition of 1 g/L K_2T causes a decrease of TA (as tartaric acid) by 1.663 • 0.4 = 0.67 g/L. As a reminder for better understanding of this calculation: The precipitation of 1.0 g/L KHT lowers TA by 0.4 g/L (Chapter 2.3). Therefore, lowering TA by 1.0 g/L requires 1.52 g/L K_2T under these conditions.

As frequently occurs under practical winemaking conditions, the KHT does not precipitate completely. To overcome this effect, one might add higher amounts of K_2T on an empirical basis. However, when in the presence of crystallization inhibitors no precipitation takes place, there will be no deacidification effect at all. This is in contrast to what happens when potassium carbonates are applied. The reason is easy to understand: When potassium carbonates are added to wine, their carbonate anion escapes as CO_2 without any effect on the deacidification reaction. When K_2T is added, the anion – tartrate – is preserved as long as no KHT drops out.

The original interest in the use of dipotassium tartrate lies in the fact that it produces additional amounts of KHT. As a result, there will be a higher supersaturation of KHT expected to facilitate the precipitation of KHT naturally occurring in wine, according to what is called the "contact process" (Chapter 11.2).

Application rates of potassium salts - summary

For a fast overview, table 5 provides the amounts of the various potassium compounds required to lower TA by 1.0 g/L depending on whether the potassium precipitates or remains in the wine.

Table 5: Application rates (g/L) of the various potassium (K^+) compounds (100 % purity) used for deacidification by 1.0 g/L TA (as tartaric acid).		
	K^+ precipitates entirely	K^+ remains entirely in wine
K_2CO_3	0.460	0.921
$KHCO_3$	0.667	1.334
KOH	0.374	0.748
K_2T	1.520	∞

8.

Chemical deacidification with calcium carbonate

The use of calcium carbonate is limited to unripe harvests in cool-climate growing regions, where it is increasingly being replaced by biological deacidification
using malolactic fermentation. Its advantage is that, under certain conditions of
implementation, it can also remove some malic acid when tartaric acid is not
sufficient for the desired reduction in total acidity. This process is known as
double-salting in the simple and extended versions. Nevertheless, it is justifiably
the least popular form of chemical deacidification. The reason lies in increased
levels of unstable residual calcium which it leaves behind which must be reduced by relatively laborious measures of specific calcium tartrate crystal stabilization, unless sufficient time is available for self-stabilization. Without precise
control of tartaric acid and calcium, this process is not feasible.

8.1. Standard deacidification

What is named standard deacidification is the most traditional means of chemical deacidification based on the addition of calcium carbonate ($CaCO_3$) to must
or wine. Furthermore, it is different from other applications of $CaCO_3$ by the
fact that the $CaCO_3$ is added to the entire lot of wine instead of adding it to a
portion of the total volume.

The objective is to precipitate tartaric acid as calcium tartrate within the limits
of available tartaric acid minus 1 g/L. This means for example that when the
tartaric acid level is 3.5 g/L, only 2.5 g/L can be removed and 1.0 g/L should
remain. Thus, titratable acidity is lowered by 2.5 g/L. If one leaves less residual
tartaric acid by adding higher amounts of $CaCO_3$, the levels of residual calcium
will increase dramatically with adverse enological consequences (Chapter 8.4).

The precipitation of 1.0 g/L tartaric acid requires 0.67 g/L of $CaCO_3$. This
amount derives from the underlying reaction and mass equation:

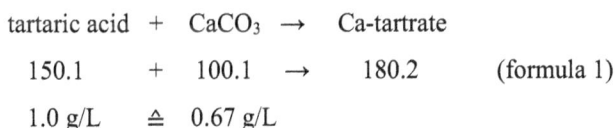

tartaric acid + $CaCO_3$ $\rightarrow$ Ca-tartrate

    150.1   +   100.1 $\rightarrow$ 180.2   (formula 1)

    1.0 g/L $\triangleq$ 0.67 g/L

In practice, a rounding up to 0.7 g/L $CaCO_3$ per 1 g/L TA is usual and justified as the lime sometimes does not entirely dissolve. Before use, its dissolution should be optimized by adding little wine to prepare a paste.

In order to facilitate the precipitation of the calcium tartrate produced, it is advisable to reverse the traditional procedure: Instead of adding the $CaCO_3$ paste to the wine, the wine is slowly added to the $CaCO_3$ previously supplied into the container. This approach improves the precipitation of the calcium tartrate produced and subsequent crystal stability (Section 8.5).

8.2. Double-salt deacidification

High-acidity musts or wines contain much more malic than tartaric acid. In these wines, there can be a need to remove more TA than there is tartaric acid. Addition of calcium ions by standard $CaCO_3$ deacidification would exhaust tartaric acid and leave elevated calcium levels causing crystal instability and a chalky aftertaste (Chapter 8.4). Thus, it is sometimes impossible to achieve the desirable decrease in TA by simple precipitation of tartaric acid with $CaCO_3$. If in such cases no physical (Chapter 9) or biological (Chapter 10) deacidification is possible or desirable, a double-salt deacidification will become indispensable. It is the only carbonate deacidification which also addresses malic acid.

The double-salt deacidification procedure has been developed in the 1960's (Münz 1960, 1961) and improved subsequently. It increases the maximum deacidification range since it allows for removal of more TA by also eliminating some malic acid. When tartaric and malic acid precipitate simultaneously with calcium ions, a double salt is generated which is called "calcium tartrate malate" or "calcium tartro-malate". Figure 14 depicts its chemical structure.

Figure 14: Calcium tartrate malate (double salt)

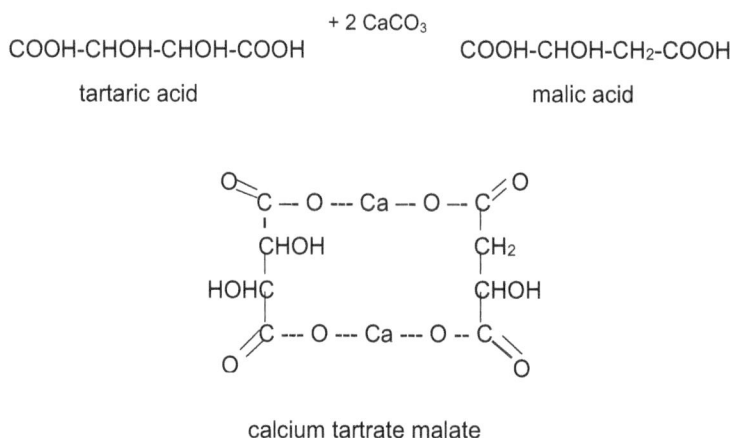

calcium tartrate malate

Two moles of acid react with two moles of $CaCO_3$. Therefore, a 1:1 relationship applies. As a result, to reduce TA by 1.0 g/L, 0.67 g/L $CaCO_3$ is required just as for standard deacidification.

The double salt contains theoretically equal molar amounts of tartaric and malic acid, but this only happens when malic acid has twice the tartaric acid level. Otherwise there will be a preferential removal of tartaric acid. Frequently the malic acid percentage of the crystal mass obtained is only 25 to 40 % (Steele and Kunkee 1978). When the double-salt procedure fails for technical reasons, the malic acid percentage is even less, while the predominant portion of the pre-cipitated crystal material is tartaric acid in the form of calcium tartrate as single salt. Furthermore, the formation of mixed crystals consisting of calcium tartrate and calcium malate is supposed (Cole and Boulton 1989).

The double-salt procedure does not require more $CaCO_3$ per 1 g/L than the standard deacidification with $CaCO_3$, but its technical implementation is sub-stantially different. Since malic acid is a weaker acid, i.e. has lower pK_A values than tartaric acid, it cannot bind to calcium ions to form a precipitate at wine pH. Hence, the pH needs to be raised to at least 4.5 to allow calcium to also bind to malic acid and to make this double salt precipitation happen.

For that purpose, double-salting is only carried out on a partial volume of the original lot. This fraction is separated and overdeacidified with the $CaCO_3$ cal-culated for the entire must/wine volume, filtered, and blended back with the un-treated residual volume. The fraction (F) is calculated as

$$F = \frac{DR \cdot 100}{TA - 2} \qquad \text{(formula 2)}$$

for musts and as

$$F = \frac{DR \cdot 100}{TA - 3} \qquad \text{(formula 3)}$$

for wines.

In these formulas, F = fraction in % of the total volume to deacidify, TA = ti-tratable acidity in g/L, and DR = deacidification range in g/L TA.

The terms -2 and -3 are applied as empirical correction factors. Their purpose is to make sure that there is enough tartaric acid in the calculated fraction, and to limit carry-over of dissolved, residual calcium from the fraction into the un-treated volume. Therefore, the fraction of wine should be somewhat larger than the fraction of juice.

The maximum deacidification range DR_{max} is calculated for musts as

$$DR_{max} = \frac{(TH_2 - 1) \cdot TA - 2)}{(TA - 2) - TH_2} \qquad \text{(formula 4)}$$

and for wines as

$$DR_{max} = \frac{TH_2 - 1) \cdot (TA - 3)}{(TA - 3) - TH_2} \qquad \text{(formula 5)}$$

Hereby, TH_2 = tartaric acid in g/L and 1 = the residual tartaric acid that has to remain in g/L. Alternatively, 1.0 g/L for residual tartaric acid can be replaced by different values ranging from 1.0 to 1.5 g/L. After blending the fraction with the untreated residual volume, residual tartaric acid should be at least 0.5 g/L in the blend.

Example: A must has 12.5 g/L TA and 3.2 g/L tartaric acid. According to formula 4, the maximum possible deacidification range DR_{max} is

$$DR_{max} = \frac{(3.2 - 1) \cdot (12.5 - 2)}{(12.5 - 2) - 3.2} = 3.16 \text{ g/L}$$

For lack of tartaric acid this must can only be deacidified by 12.5 – 3.16 = 9.34 g/L final TA.

In this case, the fraction F is according to formula 2

$$F = \frac{3.16 \cdot 100}{12.5 - 2} = 30.1 \%$$

and the amount of $CaCO_3$ required amounts to 3.16 • 0.7 = 2.2 g per liter total volume (formula 1).

Mathematics for calculating wine and $CaCO_3$ additions under stress can be tricky. Therefore, spreadsheet applications to streamline calculations are available online on various sites. Many of them do not show the correction factors. As a result, less malic acid is taken out and more calcium remains in the wine.

It must be pointed out that practitioners sometimes use even more divergent formulas for the calculation of the fraction and maximum deacidification range. They lead to different levels of residual tartaric acid and residual calcium, while the final TA looked for is always achieved. How far these differences are relevant or overlapped by other effects depends essentially on technical details during the technical conduct of double salting.

8. Chemical Deacidification with Calcium Carbonate

Double salting hands-on – step by step

- Measure the wine/must volume.

- Calculate the exact amounts of $CaCO_3$ and fraction needed. A general rule says that too small a fraction volume is less detrimental for the double salt formation than too large a fraction.

- Dissolve the $CaCO_3$ lime in some three times its volume in wine/must to prepare a paste. The lime should be fresh and reactive. If it does not produce foam in a tasting glass containing some wine, it is too slow to react and can only be used for standard deacidification. Special commercial lime brands recommended for double-salting have some double salt seeds added in order to induce and promote its crystallization. They are not absolutely necessary when the reactivity of the lime is confirmed.

- Pour the $CaCO_3$ paste into a deacidification vessel equipped with a strong propeller mixer and with headspace large enough to prevent foaming over.

- Add the fraction slowly and without interruption into the deacidification vessel that contains the lime paste over at least 30 minutes under constant and vigorous stirring using the mixer. Add it onto to the blade of the mixer. The purpose of stirring is to ensure proper reaction conditions and persistently drive out the carbon dioxide produced, which would otherwise lower the pH and impede the formation of the double salt. A low, stable wine flowrate and strong stirring make sure that reaction conditions and proper pH adjust by themselves. During the process, pH must not drop below 4.5 and should be controlled with a portable pH meter if one does not feel secure or lacks experience. If the process threatens to finish in less than 30 minutes, reduce the wine inflow rate.

- Continue stirring for another 10 minutes after the addition of the fraction to the $CaCO_3$ is finished. The reaction is completed when no more carbon dioxide is released.

- The double salt slurry is quite voluminous. The bulky structure of its hedgehog-like crystals impedes their compaction to a solid sediment. Therefore, it makes up almost half of the fraction volume at the end.

- Carrying out the double-salt method does not necessarily ensure the formation of the double salt. When the procedure fails the wine will behave as if it has been overdeacidified by the standard procedure exhausting the tartaric acid content: The TA reduction looked for will be achieved anyway, but fairly high residual calcium levels will be an embarrassing outcome (Chapter 8.4). The easiest way to monitor the success of the operation is to sample a glass of the fraction after switching off the mixer. If the crystal lees settle slowly, the process is successful. In contrast, if they settle within 10 to 15 minutes, the result has rather been a standard deacidification. That means that calcium tartrate crystals have been produced instead of double salt crystals.

– The double salt is only stable at pH higher than 4.5. Therefore, it has to be entirely removed before the fraction is recombined with the residual volume. Otherwise it would disintegrate into calcium tartrate crystals and cancel the double salting effect. Hence, the fraction has to be filtered. The filtrate obtained is immediately added to the untreated residual volume.

– The filtration should be executed without delay the same day. The high pH makes the filtrate susceptible to oxidation and microbiological degradation.

– For filtration, diatomaceous earth filters, vacuum filters, or filter presses are the best choices. Crossflow filters also look tailor-made for double-salt filtration, but they should be handled with care since there are anecdotal reports about cartridges damaged by the double-salt crystals. In any case, the filtration might be a coarse one since the crystals are bulky. Removal of the crystal mass by settling and racking is a bad improvisation causing large wine losses. Furthermore, suspended double salt crystals would get into the original lot, redissolve, release calcium, and shift calcium content to troublesome levels.

– The following details have to be taken into account for filtration: Each 1 kg of $CaCO_3$ applied will require a 3 liters volume in the filter. The double salt slurry contains approximately 50 % of wine. Diatomite filters take up an amount of double salt slurry corresponding to some 2 kg of $CaCO_3$ per 1 m^2 of filter surface area. Hence, the filter might be emptied and restarted several times when larger volumes are to be processed. On the other hand, the bulky structure of double salt crystals facilitates filtration; they act themselves as a filter aid.

– After the double salt lees have been removed by filtration, the filter cannot be used for subsequent filtration of the residual untreated wine volume without cleaning it. Otherwise the double salt crystals remaining in the filter would redissolve in the untreated portion passing through. The outcome would be a badly performed standard deacidification.

– The filtrate leaves the filter quite clean and should be immediately recombined with the original lot. Its high pH makes it prone to microbial and aroma degradation. The surplus of calcium it still contains gradually precipitates with the tartaric acid completely preserved in the untreated lot. It is essential to mix well after blending back.

– It is obvious that all the agitations and pumpings involved in the double-salt process are opposed to gentle wine treatment and detrimental to volatile aromatics of fruity white wines. A large part of them are fermentation-derived and evaporate as a function of a turbulent wine surface, temperature, and stripping out by carbon dioxide release. Hence, it is recommended to perform this kind of deacidification on musts. When this is ruled out by high potassium or pH levels (Chapter 6.4), any double-salting of wine should be executed at the lowest possible temperature to restrict aroma losses.

8.3. Extended double-salt deacidification

Many juices/wines obtained from cool climate regions, high-TA varieties, or cool seasons contain more malic than tartaric acid. This is particularly true when a large part of the tartaric acid has already dropped out as bitartrate after primary fermentation. When this happens, the double-salting described in the previous chapter runs into its limit. Its extent is constrained by the tartaric acid level. One cannot precipitate more malic acid than there is tartaric acid. Enhancing the ratio of tartaric to malic acid allows for enhancing the scope of chemical deacidification.

Such considerations led to the development and optimization of the extended double-salt deacidification in the 1980's (Würdig 1980, 1984, 1986). It is based on the same principles as the simple double-salting (Chapter 8.2). However, its fascinating feature is the possibility of unlimited deacidification regardless of the initial tartaric acid content.

Addition of tartaric acid to the entire volume of wine has often been proposed to improve the ratio of tartaric to malic acid and allow for the removal of higher amounts of malic acid. However, this approach is uneconomic and makes little sense. A major part of the tartaric acid added would spontaneously precipitate as potassium bitartrate and not be any more available for double salt deacidification.

As a better solution already established in the wine industry, the tartaric acid is added only to the fraction to be overdeacidified. This only happens after that fraction has been mixed with the $CaCO_3$. Obviously, the amount of $CaCO_3$ has to be increased by an equivalent amount corresponding to the added tartaric acid.

The fraction is calculated according to formulas 2 or 3 as for any other double-salt deacidification. The tartaric acid to add (TH_2add) is obtained by the formula

$$TH_2add = DR - TH_2 + 1 - \left[\frac{DR \cdot TH_2}{TA - 2} \right] \qquad \text{(formula 6)}$$

Whereby DR = deacidification range for musts as for wines in g/L TA, TH_2 = tartaric acid content in g/L, 1 = residual tartaric acid to be left in g/L, TA = titratable acidity in g/L, and TH_2add = tartaric acid to add in g/L related to the entire volume to deacidify.

When TH_2add yields less than 0, the deacidification range sought can be obtained by double salting without adding tartaric acid.

The amount of $CaCO_3$ to add is (DR + TH_2add) • 0.7 g/L, referring likewise to the total volume to deacidify.

Just as for any other double-salt deacidification, the fraction is pumped onto the lime paste under constant and vigorous stirring. Only after the reaction has stopped and foaming ceased, the calculated amount of tartaric acid is slowly added in small subsets into the overdeacidified fraction. Thereby, foaming will start again.

If the tartaric acid is added by mistake before or altogether with the lime, the procedure will not work out. Though the final TA would be reached, acidity would only be neutralized by calcium ions without being precipitated. The reason is easy to understand: The pH in the fraction would not increase sufficiently to ensure the formation of the double salt.

All other steps and hands-on recommendations for practical implementation of the procedure are the same as for the common double salting.

For making mathematics easier, tartaric acid to add can be replaced by at least 1.7 times its amount of a mixture containing tartaric acid plus the equivalent portion of $CaCO_3$. Mixtures of that kind are commercially available. They can even be added in surplus without affecting the outcome of the deacidification process. When they are used, the need to previously add true $CaCO_3$ is reduced to 0.7 g/L per 1.0 g/L TA to remove, according to formula 1. All other calculations are identical to those outlined in chapter 6.2 for simple double salting.

A question of workload and accurate data

Double-salting is a laborious procedure. For that reason, one prefers standard deacidification as long as there is enough tartaric acid available for achieving the final TA required. However, the correct way to deacidify is never a matter of choice.

It is important to remember that the instantaneous level of tartaric acid determines whether standard, double-salt, or extended double-salt deacidification is required. Tartaric acid figures measured several weeks ago may not remain valid since they might have decreased by bitartrate precipitation. Inaccurate tartaric acid data are one of the primary reasons of failure of the double-salt process.

On the other hand, when there is enough tartaric acid, it doesn't make any sense to perform double-salting with the sole objective of better preserving it by removing more malic acid. Such attempts are sometimes justified by the obsolete assumption that tartaric acid tastes ripe and malic acid green and harsh. This assumption lacks any scientific basis (Chapter 2.1). Any surplus tartaric acid would definitely drop out as bitartrate during further aging.

8.4. High calcium contents and their consequences

As outlined in previous chapters, 0.7 g/L $CaCO_3$ unfailingly neutralizes 1.0 g/L TA, expressed as tartaric acid. This decrease of TA occurs instantaneously upon addition of $CaCO_3$ regardless of whether the resulting calcium salts precipitate or not. However, it is assumed hereby that the $CaCO_3$ dissolves entirely. This is not guaranteed when older products are used that have been stored under damp conditions. $CaCO_3$ is hygroscopic and changes crystal structure when it absorbs moisture from the air. If it does not dissolve completely, the final TA will not be achieved as calculated.

Under optimum conditions the calcium ions added in the form of $CaCO_3$ can drop out for the most part as calcium tartrate (standard deacidification) or calcium tartro-malate (double-salt deacidification). Optimum deacidification conditions mean above all that for deacidification one only makes use of that part of tartaric acid that is more than approximately 1 g/L. It could be considered as precipitable tartaric acid, which is usable for deacidification. All calculations must be made accordingly. If calculations do not take into account 1 g/L of residual tartaric acid, calcium precipitation will inevitably be incomplete. As a result, consistently high calcium levels will remain in the wine. Those wines are considered as overdeacidified regardless of final TA achieved.

Many jurisdictions raise objections to wines containing less than 0.5 g/L of tartaric acid or to wines displaying high calcium levels expected to retroactively precipitate tartaric acid to below that threshold. OIV regulations recommend 1.0 g/L of residual tartaric acid. The basic purpose of these legal restrictions is to prevent overdeacidification.

Elevated calcium contents remain unchanged over a long time

In real life, most deacidifications using $CaCO_3$ leave increased calcium levels behind even when they are performed appropriately. The calcium incorporated into the wine does not precipitate entirely and immediately. Instead, residual calcium is subject to a slow and mostly incomplete post crystallization over several weeks and months. Thereby the initial calcium levels measured before deacidification are rarely attained again. Though a stabilization on its own is conceivable and facilitated by bentonite fining of white wines and filtration (Postel et al. 1984), it will be hardly complete until bottling. After most deacidifications with $CaCO_3$, enhanced calcium levels of 150 to 300 mg/L Ca^{++} remain in the wine. They even tend to be higher in wines obtained from rotten fruit.

As a general rule, calcium post crystallization is faster when wines are filtered and show increased (> 1 g/L) tartaric acid levels. Filtration depletes colloids inhibiting crystallization. However, in contrast to potassium bitartrate stabilization most wineries are familiar with, cold storage does not promote the crystallization of calcium tartrate but rather hampers it (Chapter 12). General measures to reduce residual calcium as early as during the conduct of deacidification are discussed in section 8.5.

Elevated calcium levels have a significant enological importance. They affect sensory quality on the palate and post-bottling crystal stability. This applies irrespective of whether they have been generated by overdeacidification, appropriately conducted deacidification, or by any other origin such as from vineyard soils.

Sensory consequences

The sensory impact of elevated calcium levels is not directly related to the TA decrease they cause. Rather they elicit an aftertaste that is described in sensory terms as sticky, chalky, floury, and grinding. Frequently, that aftertaste is not correctly identified and confounded with the burning sensation evoked by high alcohol contents, the astringency of tannins, or the scratching of volatile acidity. Those wines generally contain more than 200 mg/L of calcium, while natural calcium levels usually vary in the range of 60 to 120 mg/L Ca^{++}.

Sensory exercise 2 helps training the sensory identification of calcium.

Impact on crystal stability

When wines are bottled with elevated calcium levels, there will be a serious risk of post bottling calcium tartrate precipitation. These precipitations have crystalline structure. The corresponding crystals, however, are usually too small to be identified as such by the naked eye; they rather appear as an amorphous deposit or cloudiness.

In former times a waiting period of some six weeks was generally recommended after deacidifications with $CaCO_3$. The idea was that unstable calcium would drop out spontaneously within this time frame to an extent that would provide crystal stability. This assumption is a long way away from reality. The spontaneous precipitation of surplus calcium does not allow for any standardization. It generally takes a period of several months if indeed it ever happens before bottling (Postel et al. 1984). Thus, it is quite normal that wines after deacidification with $CaCO_3$ show unstable calcium levels of more than 200 mg/L Ca^{++} over a long period.

Identification of calcium tartrate deposits in bottled wines

The identification of calcium tartrate crystals in post-bottling deposits is relatively easy by addition of sulfuric acid 10 or 25 %. They do not dissolve in sulfuric acid but are converted into insoluble, white crystals of calcium sulfate (gypsum). In contrast, potassium bitartrate crystals dissolve under the same conditions within a couple of minutes (Klenk and Maurer 1967).

Based on this behavior, one can easily differentiate between the origins of crystal deposits. Measurement of calcium and potassium in the crystal mass provides for additional reliability. There is an even less sophisticated hands-on test based on tasting: The more common potassium bitartrate crystals display a sour taste on the palate as they represent a sour salt, while calcium tartrate crystals do not have any taste.

Trying to distinguish both forms of crystals by the naked eye is not reliable. That's why the description of their physical appearance does not make very much sense. However, it should be noted that calcium tartrate produces small crystals with an irregular, almost dust-like look, while potassium bitartrate crystals display a more visible crystalline shape. In some wines, a mix of both of them can be observed at the same time.

8.5. Promoting calcium tartrate precipitation during deacidification

The basic rules and technical measurements of promoting crystallization during deacidification are outlined in Chapter 7.5. They also apply to the precipitation of calcium tartrate.

One of the practical consequences of these physicochemical laws is that regardless of the deacidification agent ($CaCO_3$, $KHCO_3$, or K_2CO_3) used, it should be put into the deacidification vessel first and the wine then added slowly, and not the other way around. Alternatively, it is dissolved in a fraction of the total wine volume, which is overdeacidified and then blended back. For both procedures the advantage is the same: From the very beginning of the operation, the tartaric salts attain fairly high concentrations that facilitate their precipitation. There are, however, some peculiarities with regard to the use of $CaCO_3$ and subsequent calcium tartrate crystallization, which differ from potassium bitartrate precipitation:

- The formation of calcium tartrate requires tartrate anions (T^{2-}) whose concentration increases with pH (Figure 4). When the wine is added to $CaCO_3$ (instead of $CaCO_3$ to wine) or when the $CaCO_3$ is previously dissolved in a partial volume, a high pH promoting calcium tartrate precipitation is created from the start. This pH effect is more pronounced on calcium than on potassium bitartrate precipitation, which reaches its maximum at pH 4.1 and declines at higher pH levels.

  Dissolving $CaCO_3$ in a fraction is indispensable for removing some malic acid during the conduct of the double salt procedure (Section 8.2). Doing so in the context of standard deacidification (Section 8.1) has the sole purpose of making use of high pH and maximum calcium tartrate supersaturation to facilitate its precipitation without removing malic acid. From that point of view, the fractional percentage should be as low as possible. On the other hand, the fraction must contain the whole amount of tartaric acid to be removed. Under these conditions, the maximum amount of calcium tartrate crystal mass can be obtained in the fraction. Blending back with the residual volume does not require previous filtration as required for double-salting since calcium tartrate crystals will not redissolve. In the absence of filtration, there will be no removal of malic acid.

- Previous cooling of the wine additionally increases calcium tartrate supersaturation, thus promoting crystallization. A temperature around 10° C (50° F) has been recommended for deacidification with $CaCO_3$ (Friedrich and Müller 1999). However, it is important to keep in mind another basic rule underlying calcium tartrate precipitation: It is only fostered by cooling when calcium tartrate occurs at the highly supersaturated concentrations that are generated throughout deacidifications with $CaCO_3$. At low supersaturation levels existing after completion of deacidification, cooling will be counterproductive because calcium tartrate crystallization is an endothermic reaction requiring energy (Chapter 12.2).

9.

Deacidification by physical means

Physical procedures as ion exchange and electrodialysis used for acidification (Chapter 4) and potassium bitartrate stabilization (Chapter 11.2) can also be applied to deacidification after some modifications of the device setup. Since most of these procedures are based on membrane processes, they require a previous fine filtration of the wine, which is not needed for chemical deacidification. Their advantage is that in contrast to the chemical deacidification procedures using calcium or potassium carbonates, they do not leave any residual potassium or calcium in the wine. Thus, they make subsequent measures of crystal stabilization less important. Furthermore, their functioning is not specific to any acid; it does not require the presence of minimum amounts of tartaric acid. Despite these benefits, their legal approval and practical use for deacidification purposes has been restricted.

9.1. Ion exchange

For deacidification by ion exchange, acid anions of the wine are replaced by OH^- ions using a weakly basic anion exchange resin. The acid anions are bound onto the resin until depletion of its exchange capacity. As a side product, water is generated instead of carbon dioxide produced by chemical acidification with carbonates.

The resin can be added to the wine in a batch process for treatment of small wine volumes, but a continuous flow-through process in conjunction with reverse osmosis is more frequent. For that purpose, the wine is first passed through reverse osmosis splitting the wine into a colorless, flavorless permeate and a retentate fraction. Depending on the porosity of the membrane used, acids and some other small molecules pass into the permeate, while all other intrinsic wine constituents remain in the retentate. Only the permeate is passed through the ion exchange column and then recombined with the retentate. This procedure has less undesirable side effects than passing the whole wine through the exchange column.

Because of the possible release of traces of quaternary ammonium salts affecting wine sensory properties, anion exchange resins are not authorized in Europe for wine deacidification, but only for the production of concentrated rectified musts.

Two-step reverse osmosis

A two-step design of reverse osmosis combines physical and chemical features of deacidification. In the first step acids, preferentially malic acid, are concentrated in permeate I and neutralized by addition of potassium hydroxide. The neutralized permeate I thus obtained is passed in a further step through a second, identical membrane, which retains the potassium salts of the acids in the retentate up to almost 100 %. The outgoing permeate II is identical with permeate I minus acids and recombined with the starting wine. Thus, there is no need of ion exchange. Crystal stability is not affected (Ducruet et al. 2007).

9.2. Electrodialysis

While reverse osmosis applied for deacidification uses membranes that only pass undissociated acids, there are other membranes that pass only the ions. Thus, when wine is pumped between two membranes, a cation-permeable one allows diffusion of only cations, and an anion-permeable one passes anions such as tartrate and malate. The driving force is a low-voltage DC current, which makes cations gravitating to the negative pole and anions to the positive pole. The ions are drawn into a brine that is discarded. This is the working principle of electrodialysis described in more detail in section 4.2 with regard to its application for acidification. For this purpose, bipolar membranes are coupled with cationic membranes, allowing the replacement of potassium and other cation by protons (H^+).

Conversely, for deacidification, bipolar membranes are coupled with anionic membranes. The wine flows between anionic membranes and the anionic side of the bipolar ones. When the electric field is applied, the anions of the acids move to the anode, crossing the anionic membrane. Thus, they are extracted from the wine and replaced by the OH^- ions, which are produced at the junction layer of the bipolar membrane. Potassium and other cations tend to move to the cathode, but they remain in the wine compartment because they cannot cross the anionic layer of the bipolar membrane. As the process continues, wine is progressively enriched with OH^- ions, while organic acids are extracted and concentrated in the brine. As a result, an increase of pH and a decrease of TA is observed (Comuzzo and Battistutta 2018).

10.

Deacidification by biological means

This chapter is mainly about malolactic fermentation. Since yeasts are not a practicable means of acidity reduction without quality losses, it is the only way of biological deacidification. One of its advantages is that, in contrast to the methods of chemical deacidification, it leaves the cation balance of the wines unaltered, which has consequences for in-mouth sensations and simplifies crystal stabilization. Additionally, the well-known conversion of malic acid into stable lactic acid as well as the reduction of residual nutrients able to support microbial growth make the wine more stable during further storage. However, the effect of the malolactic bacteria involved in this process goes far beyond the reduction of titratable acidity. The formation of diacetyl and other by-products can have a lasting sensory impact on wines irrespective of the final acidity achieved. This impact can be modulated by the choice of specific bacteria starter cultures, inoculation strategies, and ways of overcoming limiting factors of bacterial activity.

10.1. Deacidification by malolactic fermentation

Once the alcoholic fermentation has finished, or simultaneously in some cases, most red but also some white wines undergo malolactic fermentation (MLF). In this process, malolactic bacteria (MLB) convert the diprotic malic acid into the monoprotic lactic acid and carbon dioxide with no intermediate products. It can be induced by autochthonous MLB or by inoculating with selected starter cultures. These freeze-dried bacteria cultures consist predominantly of strains of *Oenococcus oeni*, previously also known as *Leuconotoc oenos*. In more recent times, the array of suitable strains has been completed by some of the *Lactobacillus plantarum* genus.

Spontaneous malolactic fermentation

Before the use of MLF starter cultures became a common winemaking practice, MLF was often induced by contamination with the microbial populations that originated in the winery. All that was needed was to wait long enough for the ubiquitous bacteria to become active before the first post-fermentation SO_2 addition. This uninoculated MLF by 'wild' MLB is still used without any problems by some wineries in traditional Old World viticultural areas, but its outcome is highly winery-dependent and cannot be generalized. It all depends on which strain of bacteria comes to dominate in the winery.

One of the major risks of spontaneous MLF is the early onset of activity of the autochthonous bacteria strains involved. Many if not most of them are relatively resistant to low pH and low temperatures, can become active very quickly, and are able to convert residual sugars to acetic acid under anaerobic conditions. Cases of inadvertent MLF at only 8° C (46° F) and pH 3.2 are known. Obviously, some spontaneously occurring 'wild' bacterial strains are more resistant to low temperatures and low pH levels than most commercial starter cultures. If such a propensity can be observed in a winery, care must be taken to avoid sluggish fermentation and to remove the sugar in due time. If in the worst case the problem can no longer be brought under control, sterilization of the entire winery equipment with steam or hot water may be necessary. Simple disinfection is not sufficient.

Indeed, sluggish and stuck fermentations are a very common cause of elevated levels of volatile acidity. In some cases, an increase in volatile acidity of up to 0.5 g/L per day was observed under these conditions. It should be noted that volatile acidity in still fermenting young wines with residual sugar is very difficult to detect by sensory means due to masking effects. Therein lies the major risk of sluggish fermentations.

Another drawback of spontaneous MLF is the formation of elevated levels of biogenic amines, to which some people experience allergic reactions. The use of starter cultures can help to minimize this risk.

Inoculation strategies

When selected bacteria strains are used, the winemaker can chose between three inoculation strategies:

– Sequential inoculation with yeast first, whilst *Oenococcus oeni* is inoculated only at the end of alcoholic fermentation, generally after complete sugar depletion.

– Co-inoculation, whereby MLB starters are inoculated simultaneously or shortly after the beginning of alcoholic fermentation. Potential advantages of this approach, as compared to the more traditional sequential inoculation, are the reduction of the overall time required for both alcoholic and malolactic fermentation, thus enabling the winemaker to protect the wine with earlier SO_2 additions and to reduce the risk of spoilage from spoilage microorganisms such as *Acetobacter* and *Brettanomyces*.

– Sequential inoculation with MLB first, followed by inoculation with a yeast starter a few days later. This technique is not common and based on the use of *Lactobacillus plantarum* strains. In the meantime, preparations comprising mixtures of *L. plantarum* and *O. oeni* are also commercially available. Since all these strains are extremely sensitive to SO_2, it is important that SO_2 additions to must are very moderate if the addition of bacteria is planned before the start of alcoholic fermentation. For this reason, the bacterial cultures are usually inoculated only one day after the yeast. In this way, the yeast can degrade any free SO_2 that may still be present and the bacteria are no longer inhibited.

The performance of MLB starter cultures is related to the specific species, the environmental and nutritional conditions, and to microbial interactions with yeasts. Due to its high tolerance for low pH, high ethanol concentrations and scarcity of nutrients, *O. oeni* is the best known and the most used species of the *Oenococcus* genus. However, strains of *L. plantarum* are also registering increasing interest because of their fast consumption of malic acid, the suppression of the activity of spontaneous MLB populations, and their inability to produce volatile acidity through metabolism of residual sugar.

Requirements on environmental conditions

The most important condition for the activity of these bacteria is the absence of significant amounts of SO_2. They are hardly able to start MLF in the presence of any free SO_2. Therefore, MLF can only be conducted as long as the wine has not yet been subject to SO_2 additions after alcoholic fermentation. Additionally, in contrast to yeasts, MLB are also sensitive to bound SO_2, which should not exceed approximately 40 mg/L. This requires judicious handling of pre-fermentation SO_2 additions, taking into account that about half of the SO_2 added to musts is found in bound form after alcoholic fermentation, while the other half is eliminated from the system by oxidation to sulfate.

Regardless of bound SO_2, which is not related to pH, pH is still a critical parameter. pH 3.2 is commonly considered the lower limit above which the start of MLF is possible. Some special MLB strains can still initiate MLF at lower pH values if they are carefully adapted to them. However, it is easier to add 1 g/L $KHCO_3$ to raise the pH when necessary.

When sequential inoculation is used with the yeast starter added first, care must be taken to ensure that the yeast strain is compatible with MLF. This is not the case with some yeasts, which either produce too much SO_2 or consume too many nutrients, which are later no longer available for MLB. Most yeast strains produce between 10 and 30 mg/L SO_2; strains producing less than 10 mg/L SO_2 are very rare.

Nutrient requirements of MLB are more complex than those of yeasts. The measurement of yeast-assimilable nitrogen, so valuable for assessing the nutrient status of yeasts, says little or nothing about the nutrient availability to MLB. Special MLB nutrients are commercially available and they are useful when sluggish MLF occurs even though the other general requirements are fulfilled. In such cases, the addition of complex yeast nutrients has also proven helpful.

However, diammonium phosphate (DAP) is useless and not metabolized by MLB.

Most MLB strains require a minimum temperature of 18° C, but 20° C (68° F) is more advantageous. This becomes relevant when in cool-climate areas the MLB are inoculated sequentially after completion of alcoholic fermentation. Under these conditions, temperature control of the containers is usually required.

All in all, there are several limiting factors impacting start and completion of MLF. Their inhibiting effect can be potentiated by their synergistic action. On the other hand, MLF may well run smoothly if one of the factors is not fulfilled, but all the other factors are well fulfilled instead.

Impact on acidity

During MLF, the loss of one carboxyl group of malic acid causes the well-known reduction of TA. Stoichiometrically, 1.0 g of malic acid is converted into 0.67 g of lactic acid, which equals 0.56 g of TA expressed in tartaric acid (Table 4). However, microbiological processes rarely respect stoichiometric rules, and that is why one simplifies by stating that the breakdown of 1.0 g/L malic acid lowers TA by 0.5 g/L.

If a deacidification did not comprise more than just a reduction of TA to a given extent, there would be no qualitative differences between methods of deacidification. But as shown in detail in chapters 6 to 8, all methods of chemical deacidification affect the mineral cation balance in one way or another. As a result, the sensory effects rely on more than a mere removal of TA. One of the most fascinating features of MLF is that in contrast to chemical deacidification, it does not affect the cation balance. It does not leave behind any residual calcium or potassium, which might distract sensory quality and require subsequent measures of crystal stabilization.

Practical hints: Monitoring of MLF

The progress and end of MLF is monitored by measuring malic acid. This is usually done by enzymatic analysis, HPLC (high pressure liquid chromatography), or with lower accuracy by paper chromatography. MLF can be considered finished when no more than 0.2 g/L malic acid remains. Determination of the titratable acidity does not provide reliable information as to whether MLF has been completed. A crackling noise heard at the bunghole may indicate bacterial activity, but may also be due to a release of excess carbonic acid from alcoholic fermentation.

In the event of an inadvertent or spontaneous MLF, volatile acidity should also be monitored, especially if more than 2 g/L of fermentable sugars are still present. Most strains of *O. oeni* can also metabolize residual sugars after completion of malic acid degradation to produce acetic acid, i.e. volatile acidity. This is also where the risk of co-inoculation with *O. oeni* lies, especially if alcoholic fermentation gets stuck for some reason.

MLF comprises more than only acidity reduction

Apart from the sole conversion of bivalent malic acid into the less sour mono-valent lactic acid, MLF comprises much more than just lowering acidity, influencing sensory quality in a subtle, manifold, and complex way that goes far beyond the decrease of TA. The reasons for this are the myriads of side-products of bacterial metabolism that are not completely investigated. In sensory terms, diacetyl is the most important of them. The sensory changes they induce are especially appreciated in red wines, so that the importance of MLF for these wines does not require further discussion. They are topics of a large research area whose results are covered in numerous peer-reviewed articles (Versari et al. 1999, de Revel et al. 1999, Nielsen and Richelieu 1999, Liu 2002, Pozo-Bayón et al. 2005, Virdis et al. 2021).

Impact of MLF on aroma

In white wines, however, the use of MLF is controversially discussed. Irrespective of the reduction of the sour taste and the increase of body and volume, more or less pronounced changes in the varietal aroma can be observed. These changes are not always perceived as positive, depending on the intended style of wine. Therefore, a closer look at the impact of MLF on wine olfactory qualities and the enological practices that influence it is justified.

The influence of MLF on the aroma profile can hardly be standardized and varies from wine to wine. In the worst case, it is described as a decrease of the fruity aroma characteristics typical of the variety, with the occurrence of a lactic, buttery aroma component reminiscent of whey, milk or butter and due to the occurrence of diacetyl produced by MLB. For this reason, there is a tendency to limit the use of MLF in white wines to those obtained from more neutral varieties in which there is less distinct varietal aroma to be lost. Under such circumstances, a modest aroma can even be enriched in a positive way by additional components resulting from MLB metabolism.

The decrease of fruity varietal aroma attributes can be explained by four possible causes:

- The concentration of the underlying aroma compounds remains unchanged, but the intensity of their perception is reduced by the sensory masking effect of diacetyl.

- Fruit aromas are metabolized by the secondary metabolism of the bacteria involved.

- Fruit aromas are destroyed chemically by the prolonged oxidative phase associated with MLF in the absence of SO_2 and elevated temperature.

- Fruit aromas are reduced in concentration by the purely physical process of mass transfer across the wine surface (evaporation) if the container is not completely topped during MLF.

The role of diacetyl

Diacetyl is already formed by yeasts during alcoholic fermentation and may even be involved in the aroma of wines that have not undergone MLF via sensory synergisms (Martineau and Henick-Kling 1995). However, concentrations with olfactory evidence can only be found after MLF.

The conditions under which diacetyl accumulates during MLF are largely known. It is formed by the bacterial degradation of citric acid and, to a lesser extent, of pyruvate. Citric acid is only degraded towards the end of MLF when malic acid is completely depleted. Unfavorable metabolic conditions, such as low pH and low temperature, enhance the synthesis of diacetyl (Revel et al. 1989) as does insufficient bacterial mass (Martineau et al. 1995). Yeasts reduce it enzymatically to acetoin and further to 2,3-butandiol, which are sensorially inactive. Therefore, it is recommended to perform MLF in the presence of the whole biomass of suspended post-fermentation yeast lees and before any measures of clarification. The more suspended yeast lees are present, the lower the accumulation of diacetyl tends to be. SO_2 additions and filtration stop the breakdown of diacetyl at least temporarily. However, it is possible to further reduce the diacetyl level by subsequent addition of yeast lees from other wines. MLB are also capable of metabolizing at a later stage the diacetyl they have produced as long as no free SO_2 is present. In all respects, premature clarification of wines post-MLF is contrary to the desire to reduce diacetyl and the buttery odor it might convey.

Furthermore, anecdotal evidence indicates a slow decrease of excessive buttery aroma notes in filtered wines stored in stainless steel. With increasing age, wines lose the specific odor profile of MLF and become similar to that of the control that has not been subject to MLF. These observations indicate that, in addition to the described microbiological degradation, a purely chemical degradation of diacetyl is also possible.

The instantaneous net concentration resulting from synthesis and concomitant degradation of diacetyl is closely related to the intensity of its flavor. Odor threshold data of diacetyl vary in a wide concentration range from 0.2 to 5 mg/L and depend considerably on the wine matrix and personal sensitivity (Davis et al 1985, Krieger 1993, Miltenberger et al. 1994, Martineau et al. 1995). The threshold value tends to be lower in white wines than in red wines. However, levels above 5 mg/L are generally even rejected in red wines. Free SO_2 reduces the odor intensity of diacetyl, which can be restored when posterior SO_2 losses occur during aging (Bartowsky et al. 2008).

Diacetyl can contribute to aroma complexity with attributes of fresh bread, caramel and hazelnuts in the concentration range below 1 mg/L, whilst the well-known one-sided 'lactic' or 'buttery' aroma attribute only occurs at concentrations above 2 mg/L (Davis 1985, Krieger 1993, Dubois 1994, Sauvageot and Vivier 1997, Weiand 1997). The involvement of ethyl lactate in the buttery odor attribute is considered probable, but appears to be of little relevance.

In the meantime MLB strains have appeared on the market that produce significantly less diacetyl, because they lack the metabolic pathway that can degrade citric acid to diacetyl. Hence they are referred to as citrate-negative. Nevertheless, they can still produce diacetyl from pyruvate that originates from alcoholic fermentation. The content of diacetyl found in wine is therefore strongly dependent on the MLB species used as a starter for MLF (Bartowsky and Henschke 2004), though the introduction of citrate-negative strains has been a great progress. However, it must also be accepted that the citrate-negative strains available on the market to date are less reliable at low pH values than the conventional strains of *Oenococcus oeni*.

Indeed, MLF starter cultures are undergoing a very dynamic process of continuous improvement and sophistication. In the meantime, they have reached a stage of development that, in contrast to former times, does not necessarily affect varietal aromatics. Thanks to this evolution, MLF is finding increasing acceptance for deacidification of unoaked white wines from cool-climate growing regions and characterized by very specific varietal aromatics, thus replacing the traditional chemical deacidification in these areas. Its positive image of a biological procedure accelerates this trend, particularly in the light of growing interest in minimal or noninterventionist winemaking approaches. Another reason is that many winemakers feel insecure about chemical deacidification or don't really master it in the manner in which it should be done. Overall, however, the decision for or against MLF in white wine is extremely subjective and dependent on the personal definition of quality, the desired wine type, and the targeted market segment.

Potential aroma losses via storage deficiencies of white wines during MLF

The impact of MLF on aroma profile and intensity of white wines can only be definitively assessed when free SO_2 has been adjusted post-MLF. Even when diacetyl has been degraded to far below its threshold, a complete return of fruity varietal aroma as present in the control without MLF cannot always be expected. It can therefore be assumed that not only effects of sensory masking by diacetyl play a role, but also an effective loss of aroma compounds responsible for varietal aroma in white wines can occur under certain conditions of MLF.

Losses of volatile aroma attributable to oxidation or to evaporation through the wine surface have been observed when MLF was conducted in only partially filled stainless steel containers as compared to completely topped containers (Schneider 2010). Losses of this type are basically controlled by the ratio of surface area to liquid volume, expressed as cm^2/L. In this context, the height or volume of the headspace is not important. Depending on the geometry of the tank, the absence of only a few liters of wine can often result in a disproportionately large liquid surface area. For example, in horizontal tanks of cubic design with a capacity of 5,000 L, a wine surface of 3 m^2 or more can often be observed, although only a few cm of liquid height are missing before overflow. This corresponds to a specific wine surface area of 6 cm^2/L, through which volatile aroma compounds come to evaporate or to be oxidized by headspace oxygen. These effects are all the greater the higher the temperature (Schneider 2019).

However, many white wines are sensitive to heat. Therefore, the minimum temperature of 18-20° C (64-68° F) required for MLF should only be maintained until MLF is finished. Furthermore, it is a good idea to completely fill the containers immediately after the end of alcoholic fermentation. Loss of fruity varietal aromas due to improper wine storage must not be attributed to MLF as such.

MLF cannot always replace chemical deacidification

Despite the increasing prevalence of MLF in white wines from growing areas where such an approach was not common in the past, it also encounters natural limitations: Extremely high TA levels such as those resulting from unripe fruit in bad years or some borderline cool-climate sites make it difficult to use MLF as the sole means of deacidification. This also applies when pH is sufficiently high (> 3.2) to start MLF. The reasons for this are two-fold:

- High TA inevitably includes high malic acid levels that, though they are the direct substrate for MLF bacteria, are susceptible of inhibiting bacterial activity when they exceed a certain concentration threshold. As a result, MLF proceeds rather slowly in those wines, thus making it necessary to hold them at the required temperature level of around 20° C (68° F) over a long period without SO2 to prevent oxidation. This kind of storage conditions during several weeks tends to distract aroma quality of white wines when the objective is producing a kind of wine with distinctive varietal aroma.

- Low cellar temperatures and limited warming capacities frequently set practical limits on MLF on a broad scale in wineries of many cool-climate growing areas. For this reason, chemical deacidification will continue to be of importance at least for white wines in those areas, particularly in years affected by poor ripeness. The basic question is only whether this intervention will mean a strong TA reduction is preferably carried out on must or young wine, or a minor TA correction performed when the wine has reached a stage allowing for previous bench trials on a sensory basis according to sensory exercise 6.

10.2. Deacidification by special yeast strains

As outlined in section 5.2, most of the commercial yeast strains used for inoculation of grape musts have only a modest impact on TA. During alcoholic fermentation, *Saccharomyces cerevisiae* yeasts synthesize diverse acids such as succinic, acetic, and citric acid, while at the same time they degrade some malic acid. However, *Sacch. cerevisiae* is a poor malic acid metabolizer; commercial strains of *Sacch. cerevisiae* only degrade 0.3-0.7 g/L malic acid during alcoholic fermentation (Schneider 2005). Expressed in TA, the synthesis of new acids even outweighs the decay of malic acid. Hence, yeast metabolism usually increases TA by 1 to 2 g/L during alcoholic fermentation.

Impact of alcoholic fermentation on acidity is not predictable

On the other hand, winemakers are used to observing less TA after alcoholic fermentation without any malolactic fermentation (MLF) occurring. This is because the increase in TA caused by yeast metabolism is obscured by the loss of TA that occurs when potassium bitartrate precipitates during and after primary fermentation (Chapter 2.3). This is particularly true when high-TA musts are fermented. However, exceptions resulting in an increase of TA during fermentation can be observed when potassium bitartrate precipitation is limited in wines obtained from musts with low TA and low tartaric acid levels, or when succinic acid is produced by yeast in higher amounts than usual. As a result, the effect of *Sacch. cerevisiae* acid metabolism on post-fermentation TA is hardly predictable and certainly not usable for directed deacidification.

Malic acid degradation by Schizosaccharomyces yeasts

Usually, wine is produced by using *Saccharomyces* yeast strains to transform sugars into alcohol, followed by *Oenococcus* or other bacteria to perform MLF whenever biological deacidification is desired, in particular in red wines. As an alternative, the use of non-Saccharomyces yeast strains able to simultaneously perform alcoholic fermentation and malic acid degradation has been studied. Some strains of the genus *Schizosaccharomyces* are able of doing so, effectively converting malic acid into ethanol and carbon dioxide.

Complete malic acid degradation by *Schizosaccharomyces* requires slightly elevated temperatures in the range of 20-25°C (68-77° F). In white musts these temperatures can adversely affect wine quality in comparison with the use of *Sacch. cerevisiae* together with induced MLF by a selected strain of *Oenococcus oeni*. Additionally, certain key volatile aroma compounds are produced in very low concentration when fermentation is conducted by *Schizosaccharomyces*. This is supposed to be the reason for the poor sensory quality of wines produced in this way (van Rooyen and Tracey 1987). As an ultimate consequence, deacidification by *Schizosaccharomyces* could not become popular in the wine industry. Instead, malolactic fermentation by appropriate bacteria strains has become a standard enological tool kit in red wine making for a long time. It is also finding increasing acceptance for white wine deacidification.

A very special effect of deacidification is the decomposition of acetic acid in spoiled wines by a few *Saccharomyces cerevisiae* strains as well as by the non-Saccharomyces genus *Lachancea thermotolerans* (Section 5.2) under aerobic conditions. This feature opens the way to a controlled biological deacetification process of wines with high levels of volatile acidity (Vilela-Moura et al. 2008), although this procedure has not yet been used in practice.

11.

Potassium bitartrate crystal stabilization

The simple cold test in the refrigerator is often used to check bitartrate stability, but it does not provide information about the critical temperature limit above which a wine is stable or below which it is unstable in the long term. The determination of the bitartrate saturation temperature in conjunction with the conductivity seeding test gives a more precise picture. For KHT stabilization in practice, the classical chilling procedures are effective but costly. This also applies to ion exchange and electrodialysis. All these subtractive methods have in common that they remove acidity and potassium from the wine, thus markedly changing the wines' composition and making them thinner. Additive methods using crystallization inhibitors are less expensive, do not interfere in the wines' chemical composition, and maintain their sensory characteristics. However, the most effective of them in the long term, carboxymethylcellulose and potassium polyaspartate, tend to cause cloudiness in red wines.

11.1. Evaluation of potassium bitartrate stability

Young wines inherently contain potassium bitartrate (KHT) in supersaturated concentration. Supersaturation is a state of solution that contains more of the dissolved material than could be dissolved under normal circumstances. The supersaturated amounts of KHT are unstable and prone to drop out as crystals at a given moment.

Unfortunately, the state of supersaturation can last several years. Hence, the question regarding KHT crystal stability at the moment of bottling relates to all wines regardless of whether they have undergone any acidity corrections. However, any addition of tartaric acid or deacidification by means of potassium salts produces additional amounts of KHT. That is why an already stable wine can become unstable again when it is subject to acidity adjustments.

The crystallization of KHT depends essentially on the

– concentrations of tartaric acid and potassium as starting compounds,

– ion activity of tartaric acid and potassium ions,

– pH,

– alcohol content,

– temperature.

Only a very specific amount of KHT is soluble in a given wine at a given temperature. When the concentrations of potassium and tartaric acid exceed that solubility limit, the wine is supersaturated and the risk of precipitation arises.

Increasing alcohol content and decreasing temperature lower KHT solubility considerably. This explains the appearance of relatively high amounts of KHT crystals in young wines shortly after alcoholic fermentation. However, the amount of solid KHT deposits in the containers does not give any evidence of whether the wine is already stable or able to drop out further KHT. Therefore, the reliable assessment of KHT stability is of particular importance in any wine regardless of the occurrence of previous acidity corrections. It must be checked just prior to bottling when all enological interventions are finished, because any treatment causing changes in pH affects bitartrate stability.

There is no universal definition for cold stability. The effect of temperature explains why any wine can display crystal stability as long as the temperature is sufficiently high. Thus, simply asking whether a wine is stable does not take into account that reality. Rather, the question must be at which temperature the wine is stable.

Concentration product

KHT is the reaction product of potassium and bitartrate ions. The bitartrate (TH$^-$) percentage of tartaric acid depends on alcohol content and pH. It passes its maximum point at pH 4.1 (Figure 5). By measuring the concentration of potassium, total tartaric acid (TH$_2$), pH, and alcohol, one can calculate the KHT concentration product (CP$_{KHT}$) using the formula

$$CP_{KHT} = [(\text{potassium, mol/L}) \cdot (\text{tartaric acid, mol/L}) \cdot (\% \ TH^- \ \text{ion})] / 100$$

The values for % bitartrate at measured pH and alcohol levels are taken from tables published in the literature (Berg and Keefer 1958, Rhein and Kappes 1979).

The CP$_{KHT}$ values thus obtained are compared with historical, safe data for the appropriate wine type. Some of such reference data for different wine types and temperatures have been published (de Soto and Yamada 1963, Berg and Akiyoshi 1971). When the calculated CP$_{KHT}$ is higher than values considered safe, the wine is expected to precipitate KHT crystals. When it is lower, it is more likely that the wine will be stable at the corresponding temperature.

Safe values are highly variable from one wine to another because of colloidal complexing agents that influence KHT stability to a variable extent. However, the concentration product method does not account for the specific colloidal composition of any given wine. For that reason, it has been largely abandoned in wine quality control. Its use would be more reliable if wine was an ideal solution.

An excursion into the mechanisms of crystallization

Unfortunately, wine is not an ideal solution. It contains organic compounds that are responsible for holding KHT in solution about its saturation level. This is because they complex a part of the potassium and tartaric acid ions, thus decreasing their ion activity (Bertrand et al. 1978). In this way, ions behave chemically as if they were less concentrated than they really are. This effect is particularly noticeable in red wines. Rather than tannins, their anthocyanins complex a fraction of tartrate ions so strongly that it is unavailable for precipitation reactions, with another portion held in a series of complexes that break up over a longer period (Balakian and Berg 1968). This behavior makes KHT stabilization of red wines a much more difficult a task than that of white wines.

Many of the compounds responsible for complexation also interfere with crystallization, restricting the growth of crystals (Pilone and Berg 1965). This interference starts as soon as the crystal nuclei are formed at a very early stage of phase transition when the ions start moving from an unsorted liquid phase to a sorted crystal structure.

In a solution that does not yet contain any crystals, the ions collide randomly as a consequence of their chaotic thermal movement. When more than two of them collide in a short time, there is a chance of crystal nuclei being formed. This happens faster the more concentrated the solution. When the supersaturation remains preserved, crystal nuclei are able to grow into visible crystals.

Crystal growth is based on the electrostatic attraction of ions by the crystal surface. However, when the solution contains other constituents such as proteins, tannins, or pectins in truly or colloidally dissolved form, they can also be attracted and adhere onto the crystal surface. In this way crystal growth can be delayed or totally suppressed. In extreme cases crystal nuclei will be prevented from formation. Inhibition of crystallization explains why solutions can remain supersaturated.

For minimizing the impact of colloidally dissolved wine constituents on KHT crystallization, there is an interest in reducing their concentration. In winemaking, fining and especially filtration are used for that purpose. However, even after polish or sterile filtration the wine still contains colloids susceptible to hampering crystallization to an extent that depends on the individual wine. In this way it is quite possible that a wine highly supersaturated with potassium bitartrate remains stable, while another one that is only slightly supersaturated drops out crystals quite fast. This behavior rules out the assessment of KHT stability by a mathematical approach based on analytically determined concentration units. Rather, it requires evaluating KHT stability for each wine by an experimental approach. For that purpose, various methods are available.

Cold and freezing tests

These tests rely on the formation of KHT crystals in a suspect wine held at reduced temperature for a specified period of time. Due to a lack of industry standards, they are not exactly defined, and there is a large variability of holding times and temperatures recommended for this kind of tests. Frequently a filtered sample of the wine is stored during four days at a temperature between -4 and +3° C (25 to 37° F) and shaken daily. Holding the filtered sample at -4° C for three days at least is advisable. Sometimes the sample is frozen and then thawed. When base wines for sparkling wine are evaluated, 1.5 % of alcohol is added previously to mimic the alcohol increase at secondary fermentation.

The absence of crystal formation at the end of the test is a rough indicator of KHT stability. If crystalline deposits are observed to redissolve upon warming to ambient temperature, the wine is generally interpreted as being cold-stable, but the presence of any persistent crystalline precipitate is taken to indicate that the wine is cold-unstable. The results are expressed as a 'pass' if there is no permanent crystalline deposit after refrigeration, or 'fail' if crystals are present. Calcium tartrate stability (Chapter 12.1) is not covered by cold tests.

Cold and freezing tests are essentially a global measure of the KHT precipitation rate, i.e., the formation of crystal nuclei, secondary crystal growth, and subsequent precipitation. Unless seed crystals acting as nuclei are provided, precipitation over the relatively short test period is rather a measure of the wine's ability to form crystal nuclei. Only highly supersaturated wines lead to visible crystal formation under the conditions of these tests. False negative and false positive results are possible.

Ultimately, these tests are a purely qualitative assessment that shows if the wine will precipitate bitartrate at the moment the tests are performed, i.e., they inform about the current stability. However, they do not give any information upon the quantitative extent of KHT instability and upon the wine's potential to become unstable during aging. Despite these deficiencies, they are widely used because of their speed, low expense, and ease.

In contrast to current stability or instability, potential instability is a measure of the wine's potential to become unstable as the wine loses crystallization inhibitors due to chemical reactions occurring during aging, even if it does not currently precipitate crystals when chilled. The tests described in the following section refer to potential stability.

Conductivity seeding test

All conductivity tests are based on indirect measurements of potassium and tartaric acid ions. Their concentration drops, along with the conductivity, when KHT crystals form. The change of conductivity is used to estimate the wine's KHT stability at a given temperature.

For the contact seeding conductivity test, a filtered wine sample is stored in a cooling bath or double jacketed beaker and cooled down to 0 to -2° C (32 to 28° F) or another temperature believed to be relevant to the style of wine. When sample temperature is stable, conductivity (in μS) is measured while the sample is stirred. After the initial conductivity reading stabilizes, the sample is seeded with at least 4 g/L finely ground reagent grade potassium bitartrate under continuous stirring.

The KHT crystal seeds act as crystal nuclei inducing the crystallization of supersaturated KHT naturally present in the wine, which attach to these added crystals in a short period of time. This process is accompanied by a corresponding drop of the wine's electrical conductivity.

The conductivity drop is recorded at minute intervals until readings are stable after two to four hours. The difference of conductivity between the start and the end of the test is the result: If the difference is less than 30 μS in white and less than 50 μS in blush and red wines, the wine is considered stable. The more the conductivity decreases during the test, the more unstable is the wine.

The original version (Müller and Würdig 1978) of this test has been modified regarding several parameters concerning temperature, holding time, the amount of seeding crystals added, and the interpretation of the results. One of these modifications led to the development of automated devices (Angèle 1992) that use higher amounts of seeding crystals, and shorter holding times, thus meeting better the particular needs of routine control under industrial conditions. Thus, many wineries consider a wine stable when the conductivity drops are less than 5 % over a 30 minute period at 0 or 5° C (32 or 41° F). It must be emphasized that the chosen test temperature and time should be validated to improve confidence of the results.

The lack of standardized test parameters is a limitation of this test. Hence, the information it provides is restricted regardless of the technical set-up used. Relatively short test times may not allow adequate precipitation for red and some colloid-rich wines. More importantly, the results give no quantitative information nor on the extent of KHT supersaturation neither on the minimum temperature required to achieve crystal stability.

Measurement of saturation temperature

When a KHT unstable wine is heated, the KHT becomes gradually more soluble. When a certain temperature is reached, it is completely soluble. This means that the wine is just saturated with KHT without displaying any supersaturation.

The temperature at which a wine is saturated according to its composition (pH, tartaric acid, potassium, alcohol, and total electrolyte content) is called saturation temperature. It is the temperature that does not allow for precipitation of any KHT nor for dissolving more of it. At lower temperatures, the KHT moves more and more to a state of supersaturation allowing for precipitation. At higher temperatures, the wine can dissolve more KHT. The more the wine is stable, the more KHT it can dissolve. This is more meaningful the other way around: The higher the KHT saturation temperature, the more unstable the wine is.

The relevance of measuring the saturation temperature resides in the fact than any wine is KHT stable when it is warm enough. The measurement is based on the rapid dissolution of added KHT crystals at elevated temperatures, which is much faster than their precipitation in the cold. As a result, this kind of analysis can be performed at ambient temperature in a short period of time.

For practical implementation of the measurements, the wine sample is warmed up to between 20 and 25° C (68 to 77° F), stirred, and checked for conductivity (C_1). Then 4 g/L of find ground KHT crystals are added while stirring is continued. If the wine is not saturated with KHT at the adjusted temperature, added KHT crystals will dissolve until saturation is achieved. The dissolution is accompanied by an increase of conductivity, which is recorded. When the conductivity shift comes to a standstill after some 20 minutes, conductivity (C_2) and temperature (T^*) are read again. The bitartrate saturation temperature T_{sat} correlates with the extent of supersaturation and is calculated according to the formula

$$T_{sat} = T^* - [C_2 - C_1) : 29.3]$$

The quotient of 29.3 describing the interrelation between conductivity increase and T_{sat} has been determined by regression analysis (Würdig et al. 1982, 1985).

Some wines, especially tannin-rich red wines, can contain very high amounts of supersaturated KHT in a way that T_{sat} is higher than 25° C (77° F). For these wines, conducting the measurement at 30° C (86° F) has been proposed (Maujean et al. 1985), resulting in the following formula for calculating T_{sat}:

$$T_{sat} = 29.91 - [(C_2 - C_1) : 58.1]$$

Under real-world conditions, the effective saturation temperature is always lower than T_{sat} measured since colloidal compounds of wines, called protective colloids, inhibit precipitation of supersaturated KHT to a certain extent. Thus, practical KHT stability is considered as given when storage temperature does not exceed T_{sat} by more than

– 2 to 3° C for $T_{sat} > 14°$ C,

– 3 to 4° C for T_{sat} between 12 and 14° C,

– 4 to 5° C for T_{sat} between 10 and 12° C,

– 5 to 6° C for T_{sat} between 8 and 10° C.

The temperatures referred to are guidelines and should be considered with some tolerance. There are no absolute thresholds differentiating clearly between stable and unstable. Higher T_{sat} figures are acceptable for colloid-rich high quality white wines and red wines (Müller 1997a).

Some wineries choose different thresholds and consider wines as stable at re-frigerator temperature when T_{sat} is not higher than

- 15° C (59° F) for red wines,
- 12° C (54° F) for white and rosé wines,
- 10° C (50° F) for base wines intended for sparkling wine.

Combining measurement of T_{sat} with the seeding conductivity test allows for obtaining more meaningful information. It takes into account the specific prop-erties of the individual wine and leads to a mathematical figure called crystalli-zation factor (CF). The CF indicates by how many degrees Celsius T_{sat} is low-ered by the seeding conductivity test as compared to the initial T_{sat}. From that arises the following recommendations for the technical implementation of KHT stabilization (Friedrich and Görtges 2004):

- CF > 5 → stabilization using the contact process,
- CF = 2 to 5 → stabilization using metatartaric acid or CMC,
- CF = 0 to 2 → stable wine, no bitartrate stabilization.

By that way some costly and cumbersome methods of KHT cold stabilization can be used in a more targeted manner for wines really requiring them or re-placed by less expensive ones (Chapter 12.3).

11.2. Potassium bitartrate stabilization by subtractive methods

In the context of KHT stabilization, there are three fundamentally different ap-proaches to achieving bitartrate stability:

- Subtractive procedures to accelerate crystallization before bottling by cold stabilization, including KHT crystal seeding by the contact process, and the temporary concentration by reverse osmosis. Unstable KHT is removed in the crystalline form.

- Subtractive procedures of lowering potassium and tartaric acid concentra-tions by electrodialysis or ion exchange. The starting substances of KHT crystals are reduced.

- Additive procedures aiming at inhibiting the crystallization process by addi-tion of inhibitors like metatartaric acid, carboxymethylcellulose (CMC), pol-yaspartate, or mannoproteins. Such inhibitors act as protective colloids. They reduce the transfer rate of KHT from the solution to growing crystal nuclei and the speed with which the crystal surface grows. Thus, the width of the supersaturation field of KHT is increased.

This section deals with the subtractive procedures, whilst stabilization by addition of crystallization inhibitors is covered in the following one, section 9.3.

Traditional cold stabilization

Cold storage is the simplest subtractive method and the most ancient way of stabilizing wine against KHT crystallization. Traditionally, it has been done by using passive cooling during winter in cool-climate growing regions, or by chilling the wine to a temperature just above its freezing point, i.e. 0 or -4° C (32 or 25° F) for two to three weeks. Under these conditions, the solubility of KHT is reduced and its crystallization facilitated. The general need of cooling makes any kind of cold stabilization the most expensive method of bitartrate stabilization.

Unfortunately, spontaneous crystallization only works satisfactorily in wines highly supersaturated with KHT. Under these conditions, crystal nuclei are formed within a relatively short period of time. These seeds induce the crystallization of the insoluble KHT. If KHT supersaturation is low, formation of seeds is delayed excessively or does not occur at all. This means that stability is not guaranteed.

Temperature and holding time required depend on the kind of wine. Filtered dry white wines may become stable within one week, but wines containing sugar or red wines usually take more time. As a matter of fact, the cooling period required is hardly predictable and can at best be estimated by experience under specific winery conditions.

When simple cold storage is used, the following associated measures accelerate the spontaneous precipitation of bitartrate:

– previous protein stabilization of white wines by bentonite fining (Berg 1960, Berg et al. 1968),

– previous filtration of any kind of wine to remove complexing agents and colloidal compounds that would halt crystal growth and delay KHT precipitation,

– when the wine is chilled using refrigeration and insulated tanks, a fast cooling by means of the flow-through principle is more effective than a slow cooling in tanks with double jacketed cooling plates or just by mother nature,

– the wine should be agitated periodically to promote crystal growth.

During cold storage, the rate of KHT precipitation is fast at the beginning and slows down with time. The amount of visible crystalline haze or deposit does not allow for evaluating the stability actually achieved.

Following cold storage, it is important to thoroughly remove all KHT crystals, be they in suspension or sedimented as a crystalline deposit, before the temperature rises. Otherwise, they would redissolve. To avoid this, the wine must be cold filtered.

Contact process

KHT crystals are formed in two concentration-related stages: In the first stage, also referred to as the nucleation phase, crystal nuclei are generated in a time-consuming way. In the second stage, these nuclei grow into crystals. The crystallization rate at a given temperature is predominantly a function of the quantity of KHT in solution above the saturation level. It can be greatly accelerated by increasing its concentration or by adding crystalline KHT that reduces the time factor required for nuclei generation. The addition of crystalline KHT is the basis of the contact process, which has been developed to practical maturity in the 1970's (Postel and Prasch 1977, Rhein and Neradt 1977).

Compared with passive cold storage, the contact process is much faster and more reliable. Its practical implementation comprises cooling the filtered wine to 0 °C (32° F), 4 °C (39° F) or any other desired stability temperature, and addition of at least 4 g/L of powdered KHT crystals with a grain size of 0.01 to 0.2 mm. These crystals serve as nuclei that hold a large surface carrying active sites. KHT molecules in supersaturated solution migrate to the active sites driven by electrostatic attraction and are incorporated into the crystal lattice. Crystal growth begins immediately. Thus, the addition of KHT seed crystals eliminates the time-consuming nuclei induction phase and reduces the time required to stabilize the wine.

Wine constituents of high molecular weight such as proteins, polyphenols, and polysaccharides delay the migration of dissolved KHT molecules to the active sites of the seed crystals. For that reason, measures have to be taken to accelerate the mass transfer. Consequently, technical implementation of the contact process requires agitation of the wine during two to three hours. Permanent mixing using a propeller mixer is much more efficient than agitation at intervals. It prevents phase separation by sedimentation of the seed crystals. Conductivity measurements are used for process control. This procedure amounts to the implementation of the mini contact process (Section 11.1) on an industrial scale.

Crystals must be removed in the cold since they would start redissolving as soon as temperature rises. In small wineries, pad filters, diatomaceous earth filters, crossflow filters, or chamber filter presses are the common means for crystal removal. Thereby, the crystals can act as a supplementary filter aid. In large-scale wineries, a hydrocyclone unit and a centrifuge are installed upstream for crystal recovery before remaining microcrystals are removed by polish filtration. Automation engineering allows for customizing the procedure to a continuous flow-through process.

Size and purity of the crystals used for seeding are crucial for the effectiveness of the process. In white wines, the crystal material can be recovered and recycled several times. However, its effectiveness is diminished as the crystal surface grows and gets contaminated by attachment of wine colloids. This effect is partially compensated by the gain of crystal material resulting from the KHT removed from the wine. Thus, the crystal material can be recycled four times according to practical experience (Maujean et al. 1986). The effective recovery and reuse of the crystals is essential to ensure viability of the method on large wine volumes.

The more often the crystal mass is recycled, the longer must be the period of contact during the stabilization treatment (Garcia-Ruíz et al. 1995) as long as crystals are not subject to periodic grinding. Particle size is even considered to affect the KHT precipitation rate more than the level of agitation (Dunsford and Boulton 1981) and should be the smallest possible. Commercially available KHT seeding crystals have a particle size of 15 to 200 µm and look powdered.

Red wines do not allow for recycling crystals since their surface is irreversibly contaminated by tannins and anthocyanins. The same applies to wines contaminated by traces of metatartaric acid, polyaspartate, or CMC.

Several modifications and improvements of the contact process have been proposed. One of these developments is based on channeling the wine at saturation temperature through a column filled with KHT crystals that absorb crystallization inhibitor compounds before the actual contact process via batch procedure takes place (Rodriguez-Clemente et al. 1990). Another modification makes use of passing the chilled wine through a diatomite filter whose leaves are coated with diatomaceous earth and KHT crystals in order to provide crystal contact surface and crystal removal by filtration simultaneously in a continuous through-flow process (Scott et al. 1981).

However cold stabilization with or without contact processing is carried out, previous bentonite fining of white wines (Berg 1960, Berg et al. 1968) and filtration are indispensable prerequisites to reduce colloidal substances that would occupy the active sites on the crystal surface and, as a result, halt crystal growth. Fouling of crystals, frequently observed when poorly filtered wines are processed, is a crucial parameter in the performance of the contact process. Therefore, analytical control of KHT stability (Chapter 11.1) during or after the stabilization process is advisable.

Losses of quality through cold stabilization and contact process

In current times, the contact process has reached a fairly high degree of technical maturity and popularity in the wine industry. Its results are reliable and consistent. However, it has several drawbacks with regard to sensory quality when compared to additive procedures using crystallization inhibitors. By extension, these drawbacks also apply to passive cold stabilization and other methods of subtractive KHT stabilization.

It is a common feature of all these procedures that they remove KHT in surplus from the wine. As with all potassium salts, KHT carries a flavor dimension, contributing to the perception of volume and weight on the palate (Chapter 2.3). When cold stabilization removes an amount of only 1 g/L KHT, this means that 207 mg/L potassium are also withdrawn from the solution. This amount corresponds to the sensory threshold of potassium. When more potassium drops out, the wine appears thinner.

Simultaneously, there is also a loss in TA. The hedonic evaluation of the TA losses depends on the type of wine. In any case, every manner of cold stabilization results in a lowering of compounds shaping taste and mouthfeel (Martínez-Perez et al. 2020).

During cold stabilization operations, containers are rarely completely topped. This means that there is a liquid surface through which the wine picks up atmospheric oxygen from the headspace. This process is accelerated at low temperatures and even more when the surface becomes turbulent during wine agitation required for proper stabilization. It is advisable to provide a nitrogen or argon blanket, but that measure rarely ensures complete absence of oxygen in the headspace. As a consequence, there is a wine development towards oxidative aging. Its sensory effect only becomes evident later when the oxygen is bound and the wine warmed up. This effect might be innocuous to most red wines, but fruity white wines react to oxygen in a much more sensitive way.

The additional filtration required to remove the KHT crystals before temperature increases is a further potential source of oxygen uptake.

Agitation by mixing also causes a release of carbon dioxide contained in some wines. The formation of CO_2 bubbles is particularly fast and complete in the presence of KHT crystals used for seeding. The escaping gas entrains volatile aromatic compounds comparable to what happens when a wine is sparged. Once more, white wines are much more sensitive to the resulting sensory effect than red wines.

Not one of the aforementioned interventions causes a significant loss of quality as such. Their sum, however, is able to yield an overworked, tired white wine. This is why many winemakers believe that cold stabilization by seeding seriously affects the quality of premium white wines with delicate flavor. Instead, they prefer passive cold stabilization or the use of CMC, polyaspartate, or metatartaric acid. Since CMC and polyaspartate have become available for long-term bitartrate stabilization, the contact process has lost much of its former importance, at least for white wines. Lower costs are an additional reason.

Ion exchange

Crystal stabilization by ion exchange can consist of treating the wine with an exchange resin in cationic form, which is charged with hydrogen (H^+), sodium (Na^+) magnesium (Mg^{++}), or a mixture of both in order to partially replace wine potassium. Alternatively, anion exchange resin in the OH^- cycle can be used to reduce bitartrate. Although the use of cation and anion exchange has been allowed for crystal stabilization in the USA and Australia, OIV regulations only approve cation exchangers regenerated in acid (H^+) cycle and under strict regulations for that purpose.

The resin can be added to the wine in a batch process, thus enabling treatment of small wine volumes. In such a case, agitation is required until the resin capacity is exhausted. Most frequently, the treatment consists of passing the wine or a fraction of it through a column containing the resin. This is a rapid and economical way for stabilizing large volumes of wine since it allows for regeneration of the resin after exhaustion. A cost assessment has revealed that ion exchange is the least expensive technology of bitartrate stabilization.

During treatment with cationic resin, the Na^+ and / or H^+ ions from the resin are exchanged for potassium ions of the wine, leading to less KHT and the formation of sodium bitartrate and / or tartaric acid that are soluble in the wine. Calcium and magnesium are also lowered to a minor extent, depending on their initial concentration.

When the cationic resin is charged only with Na^+ or Mg^{++}, the TA of the wine is not affected, but the increase of Na^+ or Mg^{++} is considered undesirable. On the other hand, when the resin is in the H^+ form, the TA of the wine increases and the pH decreases due to the exchange of the H^+ ions from the resin for the K^+ ions from the wine. This approach is of interest for treating low TA wines that would benefit from increased TA.

When an H^+ resin is used, changes on the palate are also affected by the decrease of potassium. This effect is similar to what happens upon cold stabilization. It would be beneficial for some high-pH wines whose balance is compromised by excessively high potassium contents (Nagel et al. 1975). In most wines, however, potassium contributes to weight and volume of the wine in a positive way (Chapter 2.3). Therefore, a mixed resin charged with both H^+ and Na^+ ions is often used. Na^+ has taste characteristics comparable to those of K^+.

During the cation exchange treatment, some nonspecific absorption of anthocyanins and other phenols by the resin matrix can be observed. Sensory and chromatic consequences resulting therefrom have been reported not to be a major problem (Mira et al. 2006), although there are also opposite experiences. In any case, in young red wines with color based primarily on monomeric anthocyanins that carry a positive charge, these anthocyanins are lowered by direct cation exchange with visible consequences for wine color. This can create a problem in slightly colored wines as those obtained from Pinot noir.

Comparable with the application of ion exchange for acidification (Chapter 4.1) and deacidification (Chapter 9.1), its use for KHT stabilization can also be carried out in conjunction with reverse osmosis. Thus, only the flavorless permeate is exposed to the treatment, while the essential wine constituents responsible for quality remain in the retentate and are not affected by nonspecific adsorption onto the resin. In practice, however, a fraction of the original wine is passed through the column and blended back with the untreated aliquot.

Electrodialysis

Crystal stabilization is the primary reason why electrodialysis is used in the wine industry. In contrast to its application for acidification (Chapter 4.2) or deacidification (Chapter 9.2), it removes both cations (K^+) and anions (TH^-) responsible for crystal formation when it is used for KHT stabilization.

For technical implementation of this operation, the wine is passed through a pack of numerous parallel running cells whose partition walls consist of ion-selective membranes. The membranes allow for diffusion of cations or anions in alternate order. Thereby, they reach a separate cell where they are absorbed and transported away by an electrolyte solution running in countercurrent. The driving force in the process is a DC current.

The procedure is technically complex and associated with high investment costs, but is fast and flexible with reduced energy input, which was calculated to be eight times lower than that required for cold stabilization (Bories et al. 2011, Lasanta and Gomez 2012). Adequate membranes allow for removal not only of bitartrate and potassium ions but also of calcium ions. Depletion of ions is performed until to a concentration that does not allow for further crystal formation (Wollan 2010).

Process control is run by the mini contact process, the freezer test, or by measurement of the saturation temperature (Soares et al. 2009). Since ion selectivity of the membranes is not perfect, depletion of ions also extends to those that are not directly involved in crystal formation. In this regard, the procedure is less specific than cold stabilization. On the other hand and unlike cold stabilization, electrodialysis does not entrain colloids or significant amounts of TA, thus resulting in less flavor stripping.

11.3. Potassium bitartrate stabilization by adding crystallization inhibitors

Stabilization with crystallization inhibitors leaves excess bitartrate in the wine, but prevents it from crystallization.

Metatartaric acid

Metatartaric acid is currently the most widely used product to inhibit the crystallization of unstable KHT in wines destined for early consumption. Its use is approved in most winemaking countries up to an amount of 100 mg/L, but not allowed in wines exported to Japan. It is manufactured by heating of L-tartaric acid to approximately 160° C (320° F) at reduced pressure. Thereby, an intermolecular esterification between the carboxyl group of one tartaric acid molecule and an alcohol group of another tartaric acid unit takes place. The higher the esterification rate, the better the effectiveness against crystal precipitations. The legally required minimum esterification rate is 40 %, but some metatartaric acid products can show an esterification rate as high as 44 %.

Metatartaric acid in solution is not stable because it undergoes an hydrolysis of its ester groups, leading to a slight increase of tartaric acid in wine. Therefore, its effectiveness is limited in time, depending on the storage temperature. It is stable for two years at 10-12° C (50-54° F), for one year at 15-18° C (59-64° F), for three months at 20° C (68° F), and one month at 25° C (77° F). Therefore, it has to be dissolved in cold water at 10 to 25° C (50 to 77° F) prior to use and should not be used for wines that are to be pasteurized. After it starts decomposing, precipitation of unstable bitartrate does not occur abruptly, but rather gradually with the appearance of sparse small crystals.

As metatartaric acid is partially removed by fining agents, any fining processes should be carried out before its addition, and wines should be checked carefully for their protein stability. Any protein instability is exacerbated by metatartaric acid.

After its addition to wine, metatartaric acid causes some haze which disappears within a few days. Therefore, its addition should take place at least four days before polish filtration, allowing thus for its complete dissolution. Otherwise, the filter performance may be seriously compromised. Membrane filter cartridges can even get clogged.

Any filtration removes a fraction of the metatartaric acid and thus reduces its efficacy. Hence, it should only be added to wines after coarse filtration and ready for bottling. After its addition, only one final filtration is admissible. Filter pads adsorb some metatartaric acid until they are saturated. When they are used for final filtration during bottling, the first portion of the wine should be run in a closed loop before starting to feed the filler.

Gum arabic

Gum arabic is colloid extracted from certain trees of the acacia family. It consists of a polysaccharide composed of galactose, arabinose, and rhamnose. Though it has widespread use as an additive and protective colloid in the wine industry, it cannot be considered a safe measure of crystal stabilization. Theoretically, it should delay KHT precipitation by interrupting crystal growth, but this delay is highly variable from one wine to another and rather short term. Therefore, it has been largely abandoned in this context, though it is sometimes added in conjunction with other crystallization inhibitors.

Mannoproteins

The use of yeast-derived mannoproteins for KHT stabilization is based on the observation that traditional barrel aging of white wines on the lees enhances their mannoprotein content and gives more KHT stability. Using enzymatic hydrolysis of yeast cell membranes by ß-glucanase and subsequent ultrafiltration yielded a specific mannoprotein extract, which is commercialized for KHT stabilization. Within this extract, the compounds responsible for crystal stabilization are highly glycosylated mannoproteins of a molecular mass of 30 to 40 kDa (Moine-Ledoux et al. 1997, Moine-Ledoux and Dubourdieu 2002). Other commercialized preparations of mannoproteins used in the wine industry to enhance mouthfeel do not necessarily provide any effect on crystal stability.

Practical experience is currently limited: Mannoproteins are well able to prevent the formation of KHT crystals, but they do not impede their growth once it has begun. Additions of 15 to 25 g/hl are recommended, depending on the wine. However, previous trials with increasing amounts in conjunction with cooling tests are deemed necessary. Excessive amounts are ineffective on KHT crystallization (Steidl 2009). When extremely bitartrate unstable wines require unusually high additions, mannoproteins may even flocculate and reverse the initial stabilizing effect (Gerbaud et al. 2010). Hence, wines must be assessed to determine their suitability for a mannoprotein addition, with such an assessment involving the measurement of parameters such a wine turbidity and filterability. On account of their high price and the uncertainties associated with their application and effect, mannoproteins are not widely used for bitartrate stabilization.

Carboxymethylcellulose

The effectiveness of carboxymethylcellulose (CMC) for stabilization against KHT precipitations was already shown in the 1980's (Wucherpfennig et al. 1984, 1987, 1988), but its approval for use in wine took until 2009 in the EU and most other countries. CMC is a poorly defined cellulose derivative obtained from tree wood. It consists of glucose units whose alcohol groups are esterified by sodium acetate groups, thus making its dissolution possible. Depending on their degree of polymerization and substitution, there are various types of CMC that display different degrees of solubility and effectiveness against crystallization (Guise et al. 2014).

In the meantime, most countries have legalized the use of CMC in wine for crystal stabilization in an amount of up to 100 mg/L. In wines with a low KHT supersaturation, additions of only 2 mg/L can suffice for that purpose, but higher amounts increase its effect (Gerbaud et al. 2010). Owing to its cumbersome dissolution in wine, it is commercialized in liquid form of 5 or 10 % content.

In contrast to metatartaric acid, CMC does not disintegrate over time or under thermal influence but remains stable for a practically indefinite period of time. In wines showing a high level of supersaturation of instability, however, its effectiveness seems somewhat less than that of metatartaric acid recently added. Hence, such wines can show a gradual KHT salt precipitation even after high amounts of CMC have been added. Consequently, CMC is only recommended for KHT stabilization of wines whose saturation temperature is in the range of 13-14° C (55-57° F) (Bosso et al 2010), but discouraged when it is higher than 18 °C (64° F). In other words, it should not be relied upon to stabilize grossly cold-unstable wines. Continuous improvement of CMC effectiveness is to be expected.

Unfortunately, CMC shows less inert behavior in wine than metatartaric acid. It easily reacts with proteins and readily causes a haze when the wine has not undergone rigorous protein stabilization. This reaction also applies to metatartaric acid, but it takes place faster and more easily with CMC even when unstable proteins are present only at trace concentrations. This is especially the case in red wines, which are usually not fined with bentonite for protein stabilization. Additionally, CMC also interacts with red wine phenols in general and anthocyanins in particular. Thus five liquid CMC brands were shown to cause cloudiness soon after their addition to red wine, but no precipitation of colored matter (Pittari et al. 2018). Hence, CMC cannot be considered suitable for bitartrate stabilization of red wines, though it might work out in some rosé wines.

Even though reliable use of CMC is only possible in white wines, it is making alternatives to KHT stabilization lose importance. In contrast to cold stabilization, it does not remove potassium salts contributing to mouthfeel (Chapter 2.3), is more gentle a treatment for sensitive white wines, and less expensive. Base wines for sparkling can also be easily stabilized without impact on the future mousseux. Thus, when bottle-fermented sparkling wines are produced, formation of KHT crystals in the bottles and the technical trouble resulting therefrom at disgorging can be largely prevented.

Similarly to metatartaric acid, CMC should be added after fining and at least one week before final filtration and bottling. Otherwise filter media would be prematurely blocked and membrane cartridges totally fouled. When filter pads are used for final filtration during bottling, they should be saturated with CMC by circulating the wine in a closed loop before feeding the filler.

Potassium polyaspartate

Potassium polyaspartate (KPA) is the most recent of the bitartrate stabilization additives approved in most countries. It is a polymer of potassium L-aspartate, obtained exclusively from the amino acid L-aspartic acid. The L-aspartic acid monomer used in the process is produced by fermentation. A thermal process transforms the L-aspartic acid monomer into polysuccinimide, an insoluble compound that is subsequently hydrolyzed under controlled conditions to KPA. The product for oenological use is in the form of an odorless brown powder, which is completely soluble in water. Its average degree of polymerization is 30 units of aspartic acid, and its average molecular weight is 5 KDa (Bosso et al. 2015). Since it is a relatively small polymer, it hardly impacts filtration.

The mode of action of KPA is partially based on the fact that it has a negative charge at wine pH, which allows sequestration of the K+ cations required for the formation and growth of potassium bitartrate crystals. In white wines, its stabilizing effect was found to be superior to that of CMC, and thus of course to that of metatartaric acid, even at half of the recommended application rate (Eder et al. 2015). It red wines, it seems to cause turbidity less frequently than CMC. Since such turbidity is due to a reaction of the preparations with proteins, fining with bentonite is suggested to avoid it.

Investigating the effect of different doses of KPA on six wines (three white and three red) after six and twelve months of bottle aging, it was found that a dose of 100 mg/L was sufficient to stabilize all these wines, which remained stable by a cold test (−4° C for six days) after one year of bottle ageing. Filterability using 0.45 µm filters was not affected (Bosso et al. 2020). In another trial with 13 white, rosé, and red wines, all the KPA-treated wines passed the cold test (-

4° C for three days) after three months and were even stable when the test was extended for 20 days (Scrimgeour et al. 2020).

Studies on the long-term effect are not yet available. However, based on experience to date, an effect of at least two years can be assumed.

12. Calcium tartrate crystal stabilization

Calcium tartrate instability is an inherently rare problem and generally only occurs in wines with more than 100 mg/L calcium and a pH above 3.4, as usually obtained after deacidification with calcium carbonate. Unlike potassium bitartrate, stabilization cannot be brought about by cooling, because this actually delays the process of crystallization. Added crystallization inhibitors are also largely ineffective. The most common way of stabilization is to precipitate excess calcium by adding DL-tartaric acid or its potassium salt. It forms with calcium a salt with a solubility 10 times lower than that of the calcium salt of natural L-tartaric acid and precipitates it within a few weeks at ambient temperature. The process is controlled by measuring the calcium content.

12.1. Evaluation of calcium tartrate stability

Normal wines are calcium-stable since they rarely contain more than 100 mg/L Ca^{++}. Therefore, concerns about calcium tartrate (CaT) stability are restricted to wines that have been subject to chemical deacidification with calcium carbonate. Exceptions are scarce.

The measures described in chapter 8.5 aim at combining deacidification with calcium carbonate and CaT stabilization in one single step. However, they do not guarantee crystal stability right after deacidification. At best, they are suitable to make CaT stabilization after deacidifications less cumbersome and time consuming. One of the reasons for their limited effectiveness is that many wineries are not used to complying all the technical details their practical implementation requires. Other wineries are legitimately worried about the stresses the agitations and pumpings involved have on the aromatics of fruity white wines.

For the aforementioned reasons, one must assume that any deacidification with calcium carbonate bears a fairly high risk of elevated calcium levels, which affect post-bottling CaT crystal stability and detract from sensory quality. Hence, the evaluation of residual calcium and CaT stability has become a standard quality control measure after that kind of deacidification.

At a given temperature, CaT instability is more serious the higher the calcium and tartaric acid levels. This is easy to understand, but the relationship between temperature and concentration is strongly overlapped by pH and other factors. When pH increases, CaT solubility shows a sharp decrease. The reason is that increasing pH makes the second acid group of tartaric acid dissociate more (Figure 4), thus providing more double negatively charged tartrate anions (T^{2-}) required for the reaction with calcium cations. Therefore, post-bottling precipitations of CaT typically occurs in wines with pH > 3.4 (Klenk and Maurer 1967). Yet, this pH is not an absolute criterion for calcium stability.

Since pH, calcium, and tartaric acid interact, attempts have been made to express CaT stability in mathematical terms, according to the concentration product calculated for potassium bitartrate (Chapter 11.1). One of these mathematical terms was the product of pH multiplied by calcium content (in g/L). If it was less than 0.4, wines were considered stable in the past. In the long term, such a multiplication product proved to be inappropriate to reliably predicting CaT stability. It predicted instability in wines that remained stable for many years.

In reality, relationships are much more complex just as they are for potassium bitartrate stability. CaT stability is also affected by alcohol content, all individual acids, the amount and nature of colloidal compounds, and temperature. It is decreased by tartaric acid and improved by malic acid. Consequentially, a wine shown as calcium-stable can become unstable when malic acid disappears during malolactic fermentation or when alcohol increases during secondary fermentation in sparkling wine production.

Influence of organic complexing agents

A portion of the calcium exists in complexed forms as calcium pectate and chelates of polyuronic acids. Calcium chelated in this way does not show any ion activity and is soluble (Postel et al. 1984, Abguéguen and Boulton 1993, McKinnon et al. 1995, 1996, Viaux et al. 1996). Complexation also occurs with potassium (Chapter 11.1), but it appears more important for calcium.

Direct potentiometric measurements using Ca-selective electrodes enabled quantification of the levels of ionically active calcium (Schwedt et al. 1988, Cardwell et al. 1991). In 11 wines, levels were only 36 to 47 % of total calcium content as measured by the usual AAS (atomic absorption spectroscopy) method (Scollary 1987). This error can be overcome by appropriate dilution of the samples. The specific quantification of ionically active calcium, which is the only portion relevant for crystallization reactions, has not yet found practical application in the wine industry.

The complex interactions outlined above clearly show: The solubility products of calcium tartrate as determined in aqueous-alcoholic solutions at different pH, temperatures, and alcohol content do not apply to real wine. In wines, they are two- to sevenfold higher (Berg and Keefer 1959). Wines are able to keep more calcium tartrate in solution than water due to their calcium complexation power.

Colloids hamper crystallization

Colloidal compounds such as proteins and tannins do not only improve calcium solubility by complexation. They also delay precipitation of excessive, super-saturated CaT by hampering the formation of crystal nuclei. This is the reason why supersaturated CaT takes so much time to drop out in unstable wines. Conversely, measures reducing colloidal material such as bentonite fining and filtration shorten time requirements for nuclei formation and accelerate precipitation of surplus calcium (Postel et al. 1984).

The laws governing crystallization in wine were first intensively investigated with regard to potassium bitartrate (Chapter 11.1). Basically, they also apply to CaT, but the crystallization of the latter is subject to the influence of much more complex variables. Therefore, statements on the stability of CaT are far more difficult than on potassium bitartrate stability. Expressing stability limits in concentration units should always be regarded in a critical manner.

A further complication lies in the fact that cold and freezing tests to assess CaT stability absolutely do not work out. The reasons are explained in more detail in section 12.2. Cold tests only provide information on the stability of potassium bitartrate, the common cream of tartar. However, post-bottling precipitations of potassium bitartrate are not very common after deacidifications with calcium carbonate since they deplete most of the tartaric acid required to make these precipitations happen.

For assessing CaT stability, some of the approaches used for the evaluation of potassium bitartrate stability (Chapter 11.1) apply, but with different parameter settings. In total, three methods have become important:

Measurement of saturation temperature

A filtered sample is held at a temperature T^* of 26 to 30° C (79 to 86° F), supplied with 4 g/L finely ground CaT crystals, and mixed during approximately 20 minutes while the conductivity is recorded. CaT crystals will dissolve partially causing an increase in conductivity. The conductivity increase (ΔC, in μS) is used to calculate the saturation temperature (T_{sat}) following the formula

$$T_{sat} = T^* - (\Delta C : 4).$$

Wines are practically stable when T_{sat} is less than 20° C (68° F) (Görtges and Stocké 1987, Müller et al. 1990). For short-lived wines and a storage temperature of 8 to 12° C (46 to 54° F), a T_{sat} of 22 to 25° C (72 to 77° F) is considered as the critical threshold (Müller 1997 b).

Conductivity seeding test

The conductivity seeding test for assessing potassium bitartrate stability (Chapter 11.1) is modified in the following way:

A filtered sample is held in a cooling bath or double jacketed beaker at the desired stability temperature, for example at 5° C (41° F). Conductivity is measured when the temperature is stable, with the sample being constantly stirred. Upon seeding the sample with at least 4 g/L of finely ground CaT seed crystals, supersaturated CaT starts precipitating, while conductivity decreases until it stabilizes. If the decrease in conductivity is more than 20 µS, the wine is unstable at the chosen temperature (Görtges and Stocké 1987, Müller et al. 1990). The more it decreases, the more calcium tartrate is expected to drop out in the future.

Measurement of calcium content

Wines containing less than 100 mg/L Ca^{++} can be considered as stable on the basis of empirical data from several vintages (Otto et al. 1986). On the other hand, some wines being bottled with more than 200 mg/L Ca^{++} are able to remain stable over several years post-bottling without any conclusive explanation (Müller et al. 1990). This behavior underlines the role of colloidal compounds capable of preventing precipitation of supersaturated CaT in an unpredictable way.

Among all methods to evaluate CaT stability, the direct measurement of total calcium concentration is the most widely used despite its shortcomings. It allows for the evaluation of a large number of samples within a short period of time. It is performed most frequently by AAS (atomic absorption spectroscopy), though other methods are possible.

This way of evaluating CaT stability is also valid for the particular case of dessert wines obtained from fruit affected by noble rot, in which elevated levels of mucic acid (Chapter 1.4) frequently precipitate as a calcium salt. Such a kind of haze is not expected when total calcium concentration is less than 100 mg/L Ca^{++}.

12.2. Methods of calcium tartrate stabilization

Additions of so-called protective colloids as metatartaric acid, carboxymethylcellulose (CMC) or mannoproteins used to stabilize wines against potassium bitartrate precipitations (Chapter 11.2) show very limited effectiveness against CaT precipitations. In this context, CMC is slightly more effective than metatartaric acid (Stocké and Görtges 1989). It is able to delay precipitation of supersaturated CaT but not to prevent it (Wucherpfenning et al. 1987). Because of their poor effectiveness, none of these protective colloids has found practical application to protect wines against CaT precipitations. For that purpose, there is no other choice than decreasing the concentration of the original compounds involved, i.e. calcium and/or tartaric acid.

Cooling delays spontaneous crystallization of calcium tartrate

As a matter of principle, crystallization of unstable CaT is much slower than that of potassium bitartrate. It cannot be accelerated by cooling. In contrast to potassium bitartrate that can easily be precipitated and stabilized in the cold, the formation of initial CaT crystal nuclei is an endothermic reaction, i. e., it requires activation energy in the form of warmth. There are two contrasting effects that overlap one another: On one hand, cooling decreases CaT solubility, enhances its supersaturation, and thus facilitates its precipitation. On the other hand, it removes activation energy (Clark et al. 1988).

As a practical consequence of this conflicting behavior, spontaneous crystallization in the cold only occurs when CaT exists at a highly supersaturated state. However, a concentration range of 150 to 300 mg/L Ca^{++} frequently observed after deacidification with calcium carbonate rather represents a low supersaturation level. Under these conditions, the rate of spontaneous crystallization is highest at standard winery or ambient temperature around 15 to 20° C (59 to 68° F).

Contact process using calcium tartrate seed crystals

Unlike spontaneous crystallization during the classic cold treatment, precipitation induced by addition of CaT seed crystals works well at low temperatures. They act as crystal nuclei, which do not need to be produced through the use of energy. The cold serves only to decrease solubility.

Practical application of this procedure requires a much longer contact and stirring period than the better known contact process used for potassium bitartrate stabilization (Chapter 9.2). Time requirements amount to several days (Sudraud and Caye 1983, Abguéguen and Boulton 1993, Müller et al. 1997, Minguez and Hernández 1998). As a positive side effect, CaT seeding also induces precipitation of unstable potassium bitartrate in a way that allows for simultaneous stabilization against both kinds of crystal deposits (Viaux et al. 1996). However, this approach does not work out the other way round.

Seed crystals must be finely ground with an average particle diameter of less than 100 µm, and display the physical properties of a powder. Limited availability of adequately prepared crystal material in commercial amounts has prevented this procedure from being widely applied on an industrial scale.

Generating seed crystals from a tartaric acid-calcium carbonate mixture

Another approach to CaT stabilization makes use of the fact that the crystallization rate increases with increasing level of supersaturation. However, when a wine is already supersaturated, there is no way of increasing supersaturation any more by dissolving additional amounts of CaT. Alternatively, this can be done by adding calcium carbonate and tartaric acid in an equimolar ratio. The single compounds of that mixture dissolve and react in wine generating fine-particle CaT crystals in 'status nascendi'. The size of these particles is significantly lower than the size of commercially available, powdered CaT crystals. Thus, they can act as crystal nuclei in a perfect way. Accordingly, for a given amount of CaT generated, an almost infinitely large crystal surface is available for contact processing.

The addition of 2.0 g/L tartaric acid in conjunction with 1.34 g/L $CaCO_3$, corresponding to 2.5 g/L CaT after reaction, allowed for achieving calcium stability (< 100 mg/L Ca^{++}) within 16 hours at 10° C (50° F) (Friedrich and Müller 1999). Continuous agitation of the wine is indispensable.

Electrodialysis and ion exchange

Electrodialysis as used for potassium bitartrate stabilization (Chapter 11.2) is an increasingly established processing method in industrial-scale wineries. At ambient temperature, it facilitates lowering the concentration of the ions (calcium, potassium, and tartrate) involved in crystallization to a level precluding supersaturation and precipitation (Wollan 2010). There are not yet any reports about its performance in targeted calcium reduction.

The same applies to calcium removal by ion exchange resins in the acid cycle. Though it is possible to a certain extent depending on the kind of resin used, its effect on calcium is much less documented than on potassium.

DL-tartaric acid

Addition of DL-tartaric acid is the most common treatment to reduce elevated calcium levels. In contrast to naturally occurring L-tartaric acid, it is an equimolar mixture of both optically active forms of tartaric acid. It acts by producing an almost insoluble salt with calcium, which is called calcium DL-tartrate. The solubility of that salt is approximately ten times lower than that of calcium L-tartrate occurring in unstable wines (Maurer 1997). This explains the effectiveness of the treatment. However, its use is a challenging task for non-chemists.

Based on its low solubility, the calcium DL-tartrate generated in wine can precipitate within three weeks, though longer waiting times are required for some wines. Temperatures below 15° C (59° F) and low pH delay its precipitation as they do for calcium D-tartrate. Thus, one can sometimes observe a moderate calcium depletion up until 40 days after DL-tartaric acid addition. Therefore, reduced waiting times before bottling bear the risk of post-bottling calcium precipitations, even though waiting times are much less than for spontaneous precipitation of calcium L-tartrate. When the reaction is strongly delayed, filtration or enhancing wine temperature to 15 to 20° C (59 to 68° F) can be helpful. In cases where the wine contains traces of metatartaric acid or CMC, no calcium removal can be expected.

Elevated calcium levels requiring stabilization are also to be expected after double-salt deacidification with calcium carbonate (Chapter 8.2 and 8.3). Their decrease is much easier in the high pH area when DL-tartaric acid is added to the overdeacidified and filtered fraction instead of adding it to the completely treated wine obtained after recombining fraction and residual volume (Würdig 1988). The calcium DL-tartrate crystal mass precipitated in the overacidified fraction does not need to be removed before blending with the residual volume, because it is not able to redissolve in the lower pH area. This is an important improvement of the double-salt procedure aiming at removing in one step unstable calcium thereby generated.

However, determining the right amount of DL-tartaric acid to be added requires previous calcium measurement. This is mostly impossible in the short time span available before blending back the fraction with the residual volume after double salting. Owing to its high pH, the overdeacidified fraction is quite vulnerable to degradation and requires fast processing.

According to stoichiometry, removal of 1 mg calcium requires the addition of 3.7 mg DL-tartaric acid. In practice, the effective need of DL-tartaric acid is somewhat lower because a minor portion of calcium drops out as calcium L-tartrate during reaction time. Apparently, the calcium DL-tartrate crystals are able to act as seed crystals inducing the precipitation of calcium L-tartrate to a certain extent. As a consequence, the final calcium level achieved can be lower than calculated, or surplus DL-tartaric acid will remain in solution. In both cases, a delayed precipitation of calcium DL-tartrate can occur after bottling or blending. This risk is particularly serious when high amounts of calcium have to be removed by correspondingly high amounts of DL-tartaric acid. This is the reason why in real-world conditions, only 3.0 to 3.2 mg DL-tartaric acid are added to remove 1.0 mg calcium, leaving 100 mg/L residual calcium as a safety margin.

Helpful hints to use DL-tartaric acid

Even though calcium DL-tartrate is poorly soluble, it shows a certain solubility that strongly increases when calcium levels are low (< 100 mg/L Ca^{++}). In real wines, the solubility of calcium DL-tartrate was measured as 58 to 106 mg/L, corresponding to 12 to 22 mg/L Ca^{++} (Viaux et al. 1996). This explains why calcium DL-tartrate does not entirely precipitate when very low residual calcium levels are intended or accidentally achieved.

When wines undergoing this kind of stabilization have reached the calculated level of residual calcium of approximately 100 mg/L Ca^{++}, this does not prove that the reaction is completed and that no more DL-tartaric acid is in solution. Such an assumption is ruled out by the poor stoichiometric relationship between calcium and added DL-tartaric acid. Besides controlling the calcium target value, a supplementary heat test provides additional information: Any DL-tartaric acid still dissolved in the wine precipitates at 20° C (68° F) overnight causing a visible cloudiness. Unfortunately, there are no specific analytical tools to quantify DL-tartaric acid as such in the presence of L-tartaric acid.

Treatment with DL-tartaric acid will inevitably result in a slight acidification of the wine previously deacidified. For example, when 200 mg/L calcium are to be removed with 200 • 3 = 600 mg/L DL-tartaric acid, TA will increase by 0.6 g/L. If this TA increase is perceived as distracting on the palate, it can easily be reverted with additions of potassium carbonate or bicarbonate. Under these conditions, deacidification by 1.0 g/L TA will require 1.34 g/L $KHCO_3$. The mathematical background of this calculation is outlined in chapter 7.1. The potassium ions added therewith can cause potassium bitartrate instability in a wine that was already stable. However, this kind of instability can easily be overcome by addition of metatartaric acid, CMC, or cold stabilization.

Treatment with DL-tartaric is usually restricted to wines that have undergone deacidification with calcium carbonate, but it is also useful on dessert wines obtained from grapes affected by noble rot. The mucic acid contained in these wines easily drops out as its calcium salt when calcium levels exceed 100 mg/L Ca^{++}.

Potassium DL-tartrate

In order to avoid the acidifying effect of DL-tartaric acid, it is sometimes replaced by its neutral potassium salt, potassium DL-tartrate. Stoichiometrically, 5.65 mg of that salt are required to remove 1 mg of calcium, but in real-world conditions only 5.0 mg are added for the same reasons as mentioned for DL-tartaric acid.

Potassium DL-tartrate exchanges calcium for potassium ions. When 10 mg/L calcium are removed, potassium increases by 19.5 mg/L, while TA remains unchanged. Thus, it replaces minor, subsequent deacidifications with potassium carbonates that may become necessary after the use of DL-tartaric acid, but might require subsequent potassium bitartrate stabilization. In sensory terms, it depends on the individual wine whether the increase of potassium is rated better or worse than the acidifying effect of DL-tartaric acid additions.

The effect of both potassium DL-tartrate and DL-tartaric acid is based on reactions of the DL-tartrate anion they have in common. Therefore, kinetics, time requirements for the reaction, poor reliability of stoichiometry, and control measurements to make sure that the calcium target value is reached are identical for both stabilizing agents.

Recap on calcium stabilization with DL-tartrates

– Measure calcium content and calculate the amount to remove.

– The calcium target value should be approximately 100 mg/L Ca^{++}.

– Removing 100 mg/L Ca^{++} requires 3.2 mg/L DL-tartaric acid or 5.0 mg/L potassium DL-tartrate under real-world conditions.

– Time delay between calcium measurements and treatment must not exceed two days. Otherwise, calcium can be lowered by natural precipitation in a way that treatment agent is added in excess.

– Wines to treat must not contain any metatartaric acid, CMC, or gum arabic.

– Filtration prior to treatment and storage temperatures around 15 to 20° C (59 to 68° F) accelerates calcium removal.

- Measure calcium again to check for successful treatment. Allow for more reaction time when the calcium target value is not yet reached.

- When calcium stability is achieved, blending of the wine can cause instability again, devaluing all former stabilization measures and analytical checks.

13. Literature

Abguéguen O., Boulton R.B., 1993. The crystallization kinetics of calcium tartrate from model solutions and wines. Am. J. Enol. Vitic. 44 (1): 65-75.

Amann R., 2007. Säuren in Most und Wein. Deut. Weinmagazin 19: 12-15.

Amerine M.A., Roessler E.B., Ough C.S., 1965. Acids and the acid taste. I. The effect of pH and titratable acidity. Am. J. Enol. Vitic. 16 (1): 29-37.

Angèle L., 1992. Stabisat: Contrôle de la stabilité tartrique et gestion de la production. Revue Fr. d'Œnologie 65: 43-47.

Aragon P., Atienza J., Climent M.D., 1998. Influence of clarification, yeast type, and fermentation temperature on the organic acid and higher alcohols of Malvasia and Muscatel wines. Am. J. Enol. Vitic. 49 (2): 211-219.

Atanassov P., Triphonova P., 2001. Veränderung der Titrationsacidität und der Gehalte organischer Säuren während der alkoholischen Gärung. Mitt. Klosterneuburg 51 (4): 133-137.

Balakian S., Berg H.W., 1968. The role of polyphenols in the behavior of potassium bitartrate in red wines. Am. J. Enol. Vitic.19 (2): 91-100.

Ballester J., Mihnea M., Peyron D., Valentin D., 2013. Exploring minerality of Burgundy Chardonnay wines: a sensory approach with wine experts and trained panelists. Aust. J. Grape Wine Res., 19 (2): 140-152.

Bartowsky E.J., Henschke P.A., 2004. The "buttery" attribute of wine - diacetyl - desirabiity, spoilaage and beyond. Int. J. Food Microbiol. 96: 235-252.

Bartowski E.J., Francis I.L., Belloni J.R., Henschke P.A., 2008. Is buttery aroma perception in wines predictable from the diacetyl concentration? Aust. J. Grape Wine Res. 8 (3): 180-185.

Benito A., Calderón F., Palomero F., Benito S., 2016. Quality and composition of Airén wines fermented by sequential inoculation of Lachancea thermotolerans and Saccharomyces cerevisiae. Food Technol. Biotechnol. 54 (2): 135–144.

Berg H.W., 1960. Stabilization studies on Spanish sherry and on factors influencing KHT precipitation. Am. J. Enol. Vitic. 11 (3): 123-128.

Berg H.W., Keefer R.M., 1958. Analytical determination of tartrate stability in wine. I. Potassium bitartrate. Am. J. Enol. Vitic. 9 (4): 180-193.

Berg H.W., Keefer R.M., 1959. Analytical determination of tartrate stability in wine. II. Calcium tartrate. Am. J. Enol. Vitic. 10 (3): 105-109.

Berg H.W., de Soto R.T., Akiyoshi M., 1968. The effect of refrigeration, bentonite clarification and ion exchange on potassium behavior in wines. Am. J. Enol. Vitic. 19 (4): 208-212.

Berg H.W., Akiyoshi M., 1971. The utility of potassium bitartrate concentration product values in wine processing. Am. J. Enol. Vitic. 22 (3): 127-134.

Berg H.W., Akiyoshi M., Amerine M.A., 1979. Potassium and sodium content of California wines. Am. J. Enol. Vitic. 30 (1): 55-57.

Bertrand G.L., Carroll W.R., Foltyn E.M., 1978. Tartrate stability of wines. I. Potassium complexes with pigments, sulfate, and tartrate ions. Am. J. Enol. Vitic. 29 (1): 25-29.

Bonorden W.R., Nagel C.W., Powers J.R., 1986. The adjustment of high pH/high titratable acidity wines by ion exchange. Am. J. Enol. Vitic. 37 (2): 143-148.

Bories A., Sire Y., Bouissou D., Goulesque S., Moutounet M., Bonneaud D., Lutin F., 2011. Environmental impacts of tartaric stabilization processes for wines using electrodialysis and cold treatment. S. Afr. J. Enol. Vtic. 32 (2): 174-182.

Bosso A., Salmaso D., de Faveri E., Guaita M., Franceschi D., 2010. The use of carboxymethylcellulose for the tartaric stabilization of white wines in comparison with other enological additives. Vitis 49 (2): 95-99.

Bosso A., Panero L., Petrozziello M., Sollazzo M., Asproudi A., Motta S., Guaita,M., (2015. Use of polyaspartate as inhibitor of tartaric precipitation in wines. Food Chemistry 185: 1-6.

Bosso A., Mota S., Panero L., Lucini S., Guaita M., 2020. Use of potassium polyaspartate for stabilization of potassium bitartrate in wines: Influence on colloidal stability and interactions with other additives and enological practices. J. Food. Sci. 85 (8): 2406-2415.

Boulton R.B., 1980 a. The relationships between total acidity, titratable acidity and pH in wine. Am. J. Enol. Vitic. 31 (1): 76-80.

Boulton R.B., 1980 b. The general relationship between potassium, sodium and pH in grape juice and wine. Am. J. Enol. Vitic. 31 (2): 182-186.

Boulton R.B., 1980 c. A hypothesis for the presence, activity, and role of potassium/hydrogen, adenosine triphosphatases in grapevines. Am. J. Enol. Vitic. 31 (3): 283-287.

Boulton R.B., Singleton V.L., Bisson L.F., Kunkee R.E., 1996. Principles and practices of winemaking, Chapter 15. Chapman & Hall, New York 1996.

Burkert J., Baumann F., Hartmann M., Gessner M., 2021. Natürlich säuern. Deut. Weinbau 16-17: 34-38.

Cardwell T.J., Cattrall R.W., Mrziak R.I., Sweeny T., Robins L.M., Scollary G.R., 1991. Determination of ionized and total calcium in white wines using a calcium ion-selective electrode. Electroanalysis 3: 573-576.

Celotti E., Fiorini P., Cantoni S., Marino S., 2007. Proposal for an alternative product to acidify grape musts and wines. www.Infowine.com, Internet Journal of Viticulture and Enology, 2007, No. 2.

Chidi B.S., Bauer F.F., Rossouw D., 2018. Organic Acid Metabolism and the Impact of Fermentation Practices on Wine Acidity: A Review. S. Afr. J. Enol. Vitic. 39 (2): 1-15.

Clark J.P., Fugelsang K.C., Gump B.H., 1988. Factors affecting induced calcium precipitation from wine. Am. J. Enol. Vitic. 39 (2): 155-161.

Clauss W., Würdig G., Schormüller J., 1966. Untersuchungen über das Vorkommen und die Entstehung der Schleimsäure. II. Mitteilung: Bestimmung der Schleimsäure in Wein. Lebensm.-Untersuchung-Forschung 131 (5): 278-280.

Coelho J.M., Howe P.A., Sacks G.L., 2015. A headspace gas detection tube method to measure SO_2 in wine without disrupting SO_2 equilibria. Am. J. Enol. Vitic. 66 (3): 257-265.

Cole J., Boulton R., 1989. A study of calcium salt precipitation in solutions of malic and tartaric acid. Vitis 28: 177-190.

Comuzzo P., Battistutta F., 2018. Acidification and pH control in red wines. In: Red wine technology, p. 17-34. Morata A. (editor), Elsevier Academic Press, 2018.

CoSeteng M.Y., McLellan M.R., Downing D.L., 1989. Influence of titratable acidity and pH on intensity of sourness of citric, malic, tartaric, lactic, and acetic acids solutions and the overall acceptability of imitation apple juice. Can. Inst. Food Sci. Technol. J. 22 (1): 46-51.

Coulter A.D., Godden P.W., Pretorius I.S., 2004. Succinic acid – How it is formed, what is its effect on titratable acidity, and what factors influence its concentration in wine? Wine Ind. J. 19 (6): 16-25.

Darias-Martín J., Socas-Hernández A., Díaz-Romero C., Díaz-Díaz E., 2003. Comparative study of methods for determination of titratable acidity in wine. J. Food Comp. Anal. 16: 555-562.

Davis C.R., Wibowo D., Eschenbruch R., Lee T.H., Fleet G.H., 1985. Practical implications of malolactic fermentation. Am. J. Enol. Vitic. 36 (4): 290-301.

Delfini C., Cervetti F., 1991. Metabolic and technological factors affecting acetic acid production by yeast during alcoholic fermentation. Wein-Wissenschaft 46 (6): 142-150.

Deneulin P., Le Bras G., Le Fur Y., Gautier L., 2014. Minéralité du vin: Représentations mentales de consommateurs suisses et français. Rev. Suisse Vitic. Arboric. 46 (3): 174-180.

Dequin S., Baptista E., Barre P., 1999. Acidification of grape musts by Saccharomyces cerevisiae wine yeast strains genetically engineered to produce lactic acid. Am. J. Enol. Vitic. 50 (1): 45-50.

de Klerk J.-L.: Succinic acid production by wine yeast. Thesis, Stellenbosch University 2010.

de Loryn L.C., Petrie P.R., Hasted A.M., Johnson T.E., Collins C., Bastian S.E.P., 2014. Evaluation of sensory thresholds and perception of sodium chloride in grape juice and wine. Am. J. Enol. Vitic. 65 (1): 124-133.

de Revel G., Martin N., Pripis-Nicolau L., Lonvaud-Funel A., Bertrand A., 1999. Contribution to the knowledge of malolactic fermentation: Influence on wine aroma. J. Agric. Food Chem. 47 (10): 4003-4008.

de Soto R.T., Yamada H., 1963. Relationship of solubility products to long range tartrate stability. Am. J. Enol. Vitic. 14: 43-51.

Dubois P., 1994. Les arômes du vin et leur défaults. Part II. Revue Fr. d'Œnologie 145: 27-40.

Ducruet J., Silvestri A.-C., Jamet A., Giuliani D.A., Noilet P., 2007. Développement d'une technique de nanofiltration à deux étages permettant de diminuer la concentration en acide malique des moûts. Proceedings 8th International Symposium on wine treatment: 13-26.

Dunsford P., Boulton R., 1981. The kinetics of potassium bitartrate crystallization from table wines. I. Effect of particle size, particle surface area and agitation, Am. J. Enol Vitic. 32 (2): 100-105.

Eder R., Willach M. Strauss M., Philipp C, 2015. Efficient tartaric stabilisation of white wine with potassium polyaspartate. BIO Web of Conferences 15, 02036, https://doi.org/10.1051/bio-conf/20191502036

Edwards T.L., Singleton V.L., Boulton R., 1985. Formation of ethyl esters of tartaric acid during wine aging: Chemical and sensory effects. Am. J. Enol. Vitic. 36 (2): 118-124.

Fontoin H., Saucier C., Peissedre P.L., Glories Y., 2008. Effect of pH, ethanol and acidity on astringency and bitterness of grape seed tannin oligomers in model wine solution. Food Quality and Preference 19 (3): 286-291.

Friedrich G., Müller G., 1999. Neue Verfahren zur Entsäuerung mit Calciumcarbonat und Kalium-hydrogencarbonat. Deut. Weinmagazin 2: 15-18 und 3: 28-33.

Friedrich G., Görtges S., 2004. Weinsteinstabilisierung notwendig? Deut. Weinbau 16-17: 22-26.

García-Ruíz J.M., Alcántara R., Martín J., 1995. Effects of reutilized potassium bitartrate seeds on the stabilization of dry Sherry wine. Am. J. Enol. Vitic. 46 (4): 525-528.

Gerbaud V., Gabas N., Blouin J., Crachereau J.-C., 2010. Study of wine tartaric salt stabilization by addition of carboxymethylcellulose (CMC). Comparison with the "protective colloids" effect. J. Int. Sci. Vigne Vin 44 (3): 231-244.

Gobbi M., Comitini F.,Domizio P., Romani C., Lecioni L., Mannazzu I., Ciani M., 2013. Lachancea thermotolerans and Saccharomyces cerevisiae in simultaneous and sequential co-fermentation: A strategy to enhance acidity and improve the overall quality of wine. Food Microbiol. 33 (2): 271-281.

Görtges S., Stocké R., 1987. Minikontaktverfahren zur Beurteilung der Calciumstabilität. Weinwirtschaft-Technik 11: 19-21.

Gomes Benitez J., Grandal Delgado M.M., Diez Martin J., 1993. Study of the acidification of Sherry musts with gypsum and tartaric acid. Am. J. Enol. Vitic. 44 (4): 400-404.

Gómez J., Lasanta C., Cubillana-Aguilera L.M., Palacios-Santander J.M., Arnedo R., Casas J.A., Arroyo L., 2015. Acidification of musts in warm regions with tartaric acid and calcium sulfate at industrial scale. BIO Web of Conferences 5, 02007. https://doi.10.1051/bioconf/20150502007

Guinard J.-X., Pangborn R.M., Lewis M.J., 1986. Preliminary studies on acidity-astringency interactions in model solutions and wine. J. Sci. Food Agric. 37 (8): 811-817.

Guise R., Filipe-Ribeiro L., Nascimento D., Bessa O., Nunes F.M., Cosme F., 2014. Comparison between different types of carboxymethylcellulose and other enological additives used for white wine tartaric stabilization. Food Chem. 156: 250-257.

Harbertson J.F., Harwood E.D., 2009. Partitioning of potassium during commercial-scale red wine fermentations and model wine extraction. Am. J. Enol. Vitic. 60 (1): 43-49.

Heymann H., Hopfer H., Bershaw D., 2014. An exploration of the perception of minerality in white wines by projective mapping and descriptive analysis. J. Sens. Studies 29 (1): 1-13.

Höchli U., 1997. Acides et substances responsables du pouvoir tampon des moûts et des vins. J. Intern. Sci. Vigne Vin 31 (3): 139-150.

Howe P.A., Worobo R., Sacks G.L., 2018. Conventional measurements of sulfur dioxide (SO_2) in red wine overestimate SO_2 antimicrobial activity. Am. J. Enol. Vitic. 69 (3): 210-220.

Johanningsmeier S.D., McFeeters R.E., Drake M., 2005. A hypothesis for the chemical basis for perception of sour taste. J. Food Science 70 (2): R44-R48.

Klenk E., Maurer R., 1967. Das Erkennen von Calcium-Ausscheidungen im Wein. Weinberg & Keller 14: 361-367.

Klenk E., Maurer R., 1969: Beitrag zur Lösung des Calcium-Problems bei Qualitätswein. Weinberg & Keller 16, 299-313.

Kodur, S., 2011. Effects of juice pH and potassium on juice and wine quality, and regulation of potassium in grapevines through rootstocks (Vitis): a short review. Vitis 50 (1): 1-6.

Kontoudakis N., Esteruelas M., Fort F., Canals J.M., Zamora F., 2011. Use of unripe grapes harvested during cluster thinning as a method for reducing alcohol content and pH of wine. Aust. J. Grape Wine Res. 17 (2): 230-238.

Krieger S., 1993. Aromabeeinflussung durch den BSA. Deut. Weinbau 12: 20-23.

Lasanta C., Gomez J.G., 2012: Tartrate stabilization of wines. Trends Food Sci. Technol. 28: 52-59.

Lawless H.T., Horne J., Giasi P., 1996. Astringency of organic acids is related to pH. Chem. Senses, 21: 397-403.

Liu S.-Q., 2002. Malolactic fermentation in wine – beyond deacidification. J. Applied Microbiology 92 (6), 589-601.

Maltman A., 2013. Minerality in wine: a geological perspective. J. Wine Res. 24 (3): 169-181.

Martineau B. and Henick-Kling T., 1995. Formation and degradation of diacetyl in wine during alcoholic fermentation with S. cerevisiae strain EC 1118 and malolactic fermentation with Leuconostoc oenos strain MCW. Am. J. Enol. Vitic. 46 (4): 442-448.

Martineau B., Henick-Kling T., Acree T., 1995. Reassessment of the influence of malolactic fermentation on the concentration of diacetyl in wines. Am. J. Enol. Vitic. 46 (3): 385-388.

Martínez-Pérez M.P., Bautista-Ortín A.B., Durant V., Gómez-Plaza E., 2020. Evaluating alternatives to cold stabilization in wineries: The use of carboxymethyl cellulose, potassium polyaspartate, electrodialysis, and ion exchange resins. Foods 9 (9):1275. doi: 10.3390/foods9091275.

Maujean A., Vallée D., Sausy L., 1986. Influence de la granulométrie des cristaux de tartres de contact et des traitements de collages sur la cinétique de cristallisation du bitartrate de potassium dans les vins blancs. Revue Fr. d'Œnologie – Cahier scientifique 104 (3): 34-41.

Maurer R., 1977. Verhinderung von Calcium-Ausscheidung im Wein durch Behandlung mit DL-Weinsäure. Weinwirtschaft 41: 1147-1152.

McKinnon A.J., Scollary G.R., Solomon D.H., Williams P.J., 1995. The influence of wine components on the spontaneous precipitation of calcium L(+)-tartrate in a model wine solution. Am. J. Enol. Vitic. 46 (4): 509-517.

McKinnon A.J., Williams P.J., Scollary G.R., 1996. Influence of uronic acids on the spontaneous precipitation of calcium L(+)-tartrate in a model solution. J. Agric. Food Chem. 44 (6), 1382-1386.

Mendes Ferreira A., Mendes-Faia A., 2020. The role of yeasts and lactic acid bacteria on the metabolism of organic acids during winemaking. Foods 1, 1231, https://doi.10.3390/foods9091231

Miltenberger R., Stumpf C., Köhler H., Gessner M., 1994. Säureabbau mit Bakterienkulturen. Deut. Weinmagazin 27: 22-26.

Mínguez S., Hernández P., 1998. Tartaric stabilization of red, rosé, and white wines with L(+)-calcium tartrate crystal seeding. Am. J. Enol. Vitic. 49 (2): 177-182.

Mira H., Leite P, Ricardo-da-Silva M., Curvelo-Garcia A.S., 2006. Use of exchange resins for tartrate wine stabilization. J. Int. Sci. Vigne Vin 40 (4): 223-246.

Moine-Ledoux V., Perrin A., Paladin I., Dubourdieu D., 1997. Premiers résultats de stabilisation tartrique des vins par addition de mannoprotéines purifiées (Mannostab™). J. Int. Sci. Vigne Vin 31 (1): 23-31.

Moine-Ledoux V., Dubourdieu D., 2002. Rôle des mannoprotéines de levures vis-à-vis de la stabilisation tartrique des vins. Bull. de l'OIV 857-858: 471-482.

Mpelasoka B.S., Schachtmann D.P., Treeby M.T., Thomas M.R., 2003. A review of potassium nutrition in grapevines with special emphasis on berry accumulation. Aust. J. Grape Wine Res. 9 (3): 154-168.

Müller T., 1997 a. Kristallstabilität der Weine, Teil 1. Deut. Weinmagazin 03: 17-21.

Müller T. 1997 b. Kristallstabilität der Weine, Teil 2. Deut. Weinmagazin 04: 12-14.

Müller T., Würdig G., 1978. Das Minikontaktverfahren – ein einfacher Test zur Prüfung auf Weinsteinstabilität. Weinwirtschaft 114: 857-861.

Müller T., Würdig G., Scholten G., Friedrich G., 1990. Bestimmung der Calciumtartrat-Sättigungstemperatur von Weinen durch Leitfähigkeitsmessung. Mitt. Klosterneuburg 40 (4): 158-168.

Müller T., Scholten G., Friedrich G., 1997. Stabilisierung von Weinen durch einen Kontaktprozess mit Calciumtartrat. Mitt. Klosterneuburg 47 (3): 74-84.

Münz T., 1960. Die Bildung des Ca-Doppelsalzes der Wein- und Äpfelsäure, die Möglichkeiten seiner Fällung durch CaCO$_3$ im Most. Weinberg & Keller 7: 239-247.

Münz, T., 1961. Methoden zur praktischen Fällung der Wein- und Äpfelsäure als Ca-Doppelsalz. Weinberg & Keller 8: 155-158.

Nagel C.W., Tamis L., Johnson G., Carter H., 1975. Investigation of methods for adjusting the acidity of wines. Am. J. Enol. Vitic. 26 (1). 12-17.

Nielsen J.C., Richelieu M., 1999. Control of flavor development in wine during and after malolactic fermentation by Oenococcus oeni. Appl. Environ. Microbiol. 65 (2): 740-745.

Otto K., Wucherpfennig K., Plöcker R., 1986. Calciumstabilisierung mit neutralem Kalium-DL-Tartrat. Weinwirtschaft-Technik 12, 494-499.

Pangborn R.M., 1963. Relative taste intensities of selected sugars and organic acids. J. Food Sci. 28: 726-733.

Peynaud É.: The taste of wine. The Wine Appreciation Guilt, San Francisco 1987.

Pilone B.F., Berg H.W., 1965. Some factors affecting tartrate stability in wine. Am. J. Enol. Vitic. 16 (4): 195-211.

Pittari E., Catarino S., Andrade M.C., Ricardo-da-Silva J., 2018. Preliminary results on tartaric stabilization of red wine by adding different carboxymethylcelluloses. Ciência Téc. Vitiv. 33(1): 47-57.

Plane R.A., Mattick L.R., L.D. Weirs, 1980. An acidity index for the taste of wines. Am. J. Enol. Vitic. 31 (3): 265-268.

Postel W., Prasch E., 1977. Das Kontaktverfahren, eine neue Möglichkeit zur Weinsteinstabilisierung. Weinwirtschaft 32: 866-878.

Postel W., Meier B., Graf T., 1984. Untersuchungen zur Kristallisationskinetik von Calciumtartrat in Wein. Mitt. Klosterneuburg 34: 102-107.

Pozo-Bayón M.A. et al., 2005. Wine volatile amino acid composition after malolactic fermentation: Effect of Oenococcus oeni and Lactobacillus plantarum starter cultures. J. Agric. Food Chem. 53 (22): 8729-8735.

Revel G., Bertrand A., Lonvaud-Funel A., 1989. Synthèse des substances acétoiniques par Leuconostoc oenos. Réduction du diacétyle. J. Int. Sci. Vigne Vin 23, 1, 39-45.

Rhein O., Neradt F., 1979. Tartrate stabilization by the contact process. Am. J. Enol. Vitic. 30 (4): 265-271.

Rhein O., Kappes, W., 1979. Weinstein-Berechnungen. Weinwirtschaft 115: 227-236.

Rayess Y.E., Mietton-Peuchot M., 2015. Membrane technologies in wine industry; an overview. Crit. Rev. Food Sci. Nutr. 56 (12): 2005-2020.

Rozoy E. et al., 2013. Développement d'un procédé de désacidification des vins par électrodialyse à membrane bipolaire: Étude de la faisabilité à l'échelle laboratoire. Bull. de l'OIV 86 (986-988): 187-208.

Rubico S.M., McDaniel M.R., 1992. Sensory evaluation of acids by free choice. Chemical Senses 17 (3): 273-289.

Sauvageot F., Vivier P., 1997: Effects of malolactic fermentation on sensory properties of four Burgundy wines. Am. J. Enol. Vitic. 48 (2): 187-192.

Schaller A., Paul F., 1958. Über die Aciditätsverhältnisse bei Getränken, I: Die potentiometrische Bestimmung der Titrationsacidität. Mitt. Klosterneuburg, Serie A, 8: 81-93.

Schaller A. und Paul F., 1959. Über die Aciditätsverhältnisse bei Getränken, II: Die Bestimmung der Gesamtacidität durch potentiometrische Titration nach Kationenaustausch. Weinberg & Keller 6: 379-386.

Schneider V., 1998. Kalium: Sensorische Bedeutung und önologische Differenzierung. Winzer-Zeitschrift 7: 36-39.

Schneider V., 2005. Einfluss von Hefe und Gärung auf die Säure. Winzer-Zeitschrift 11: 34-36.

Schneider V., 2010. Veränderung des Aromas durch BSA in Weißweinen. Winzer-Zeitschrift 10: 34-36.

Schneider V., 2019. White wine enology: Optimizing shelf life and flavor stability of unoaked white wines. Board and Bench Publishing, San Francisco 2019.

Schneider V. and Tracey M., 2021. Red wine enology: Tannin and redox management in red wines, Chapter 5. Board and Bench Publishing, San Francisco 2021.

Schonenberger P., Baumann I., Jaquerod A., Ducruet J., 2014. Membrane contactor: A nondispersive and precise method to control CO_2 and O_2 concentrations in wine. Am. J. Enol. Vitic. 65 (4): 510-513.

Schwedt G., Schweizer A., Hendrich G., 1988. Quantifying analysis of calcium species in wine from the point of view of calcium stability. Z. Lebensm.-Unters.-Forschung 187: 229-234.

Scollary G., 1987. Free and bound calcium content in wine: Possible monitoring of protein haze formation. Austr. & NZ. Grapegrower & Winemaker 278: 25-26.

Scott R.S., Anders T.G., Hums N., 1981. Rapid cold stabilization of wine by filtration. Am. J. Enol. Vitic. 32 (2): 138-143.

Scrimgeour N., Almond T., Wilkes E., 2020. Is KPA the magic bullet for tartrate instability in wines? Aust. & NZ. Grapegrower & Winemaker 675: 68-70.

Soares P.A.M.H., Geraldes V., Fernandes C., Cameiro dos Santos P., de Pinho M.N., 2009. Wine tartaric stabilization by electrodialysis: Prediction of required deionization degree. Am. J. Enol. Vitic. 60 (2): 183-188.

Sowalsky R.A. and Noble A.C., 1998. Comparison of the effects of concentration, pH and anion species on astringency and sourness of organic acids. Chem. Senses 23: 343-349.

Steele J.T., Kunkee R., 1978. Deacidification of musts from the western United States by the calcium double-salt precipitation process. Am. J. Enol. Vitic. 29 (3): 153-160.

Steidl R., 2009. Mannoproteine – das Ende des Weinsteins? Der Winzer 07: 17-19.

Sudraud R., Caye J., 1983. Élimination du calcium du vin par le procédé par contact utilisant du tartrate neutre de calcium. Revue Fr. d'Œnologie 91: 19-22.

Usseglio-Tomasset L. Bossa PD., 1978. Determinazione delle constanti di dissociazione dei principali acidi del vino in soluzioni idroalcoliche di interesse enologico. Rivista Vitic. Enol. 31 (9): 380-403.

Versari A., Parpinello G.P., Cattaneo M., 1999. Leuconostoc oenos and malolactic fermentation in wine: a review. J. Ind. Microbiol. Biot. 23 (6): 447-455.

Viaux L., Devaux S., Renaud S., Robillard B., 1996. Comparaison des stabilisations à l'acide tartrique racémique et au tartrate neutre de calcium dans des vins de Champagne. Étude de la solubilité du tartrate de calcium racémique. Bull. de l'OIV, 781-782: 237-249.

Vilela-Moura A., Schuller D., Mendes-Faia A., Côrte-Real M., 2008. Reduction of volatile acidity of wines by selected yeast strains. Appl. Microbiol. Biotechnol. 80, 881, https://doi.org/10.1007/s00253-008-1616-x

van Rooyen T.J., Tracey R.P., 1987. Biological deacidification of musts induced by yeasts or malolactic bacteria and the effect on wine quality. S. Afr. J. Enol. Vitic. 8 (2): 60-69.

von Nida E. und Fischer U., 1999 a. Problemfeld Säure. Deut. Weinmagazin 9: 32-36.

von Nida E. und Fischer U., 1999 b. Säuremanagement, Teil II: Genuss ohne Reue. Deut. Weinmagazin 10: 28-33.

Virdis C., Sumby K., Bartowsky E. Jiranek V., 2021. Lactic acid bacteria in wine: Technological Advances and evaluation of their functional role. Front. Microbiol. 11: 612118. https://doi.10.3389/fmicb.2020.612118

Walker R.R., Blackmore D.H., 2012. Potassium concentration and pH interrelationships in grape juice and wine of Chardonnay and Shiraz from a range of rootstocks in different environments. Aust. J. Grape Wine Res. 18 (2): 183–193.

Weiand J.: Starterkulturen im Vergleich. Deut. Weinmagazin 21: 18-23.

Wollan D., 2010. Electrodialysis – new technology for rapid, efficient, reliable tartrate stabilization. Austral. & NZ Grapegrower & Winemaker, 557: 81-85.

Wong G., Caputi A., 1966. A new indicator for total acid determination in wines. Am. J. Enol. Vitic. 17 (3): 174-177.

Wucherpfennig K., Dietrich H., Götz W., Rötz S., 1984. Einfluss von Kolloiden auf die Weinstein-kristallisation unter besonderer Berücksichtigung der Weinsteinstabilisierung durch Carboxyme-thylcellulose. Weinwirtschaft-Technik 01: 13-23.

Wucherpfennig K., Dietrich H., Otto K., 1987. Neuentwicklung zur Stabilisierung von Wein gegen kristalline Ausscheidungen. Weinwissenschaft 04: 241-265.

Wucherpfennig K., Otto K., Kern U., 1988. Praktische Anwendung von Carboxymethylcellulose. Weinwirtschaft-Technik 4/5: 13-19.

Würdig G., 1980. Versuche zur Doppelsalzentsäuerung mit Weinsäurezusatz. Weinwirtschaft 116: 1319-1320.

Würdig G., 1984. Doppelsalzentsäuerung mit Weinsäurezusatz. Weinwirtschaft 120: 102-104.

Würdig G., 1986. Die erweiterte Doppelsalzentsäuerung mittels Weinsäure. Weinwirtschaft-Technik 10: 428-430.

Würdig G., 1988. Schnellstabilisierung doppelsalzentsäuerter Weine. Weinwirtschaft-Technik 03: 11-14.

Würdig G., Müller T., Friedrich G., 1982. Méthode pour caractériser la stabilité du vin vis-à-vis du tartre par détermination de la température de saturation. Bull. OIV 613: 220-229.

Würdig G., Müller T., Friedrich G., 1985. Untersuchungen zur Weinsteinstabilität. - 3. Mitteilung: Bestimmung der Weinstein-Sättigungstemperatur durch verbesserte Leitfähigkeitsmessung. Weinwirtschaft-Technik 06: 188-191.

INDEX

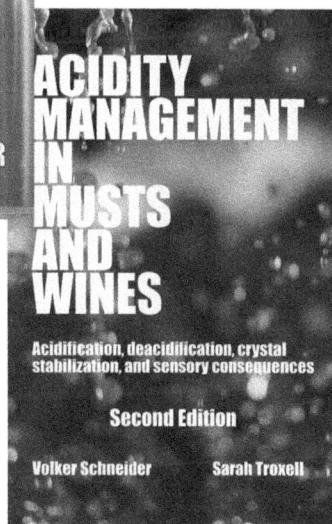